5-

Sharing the Sky

A Parent's and Teacher's Guide to Astronomy

Sharing the Sky

A Parent's and Teacher's Guide to Astronomy

David H. Levy

Larry A. Lebofsky
Nancy R. Lebofsky

PLENUM TRADE • NEW YORK AND LONDON

Library of Congress Cataloging-in-Publication Data

On file

Unless otherwise noted, all illustrations are credited to Project ARTIST. The star charts were created by the authors using Epoch 2000 Sky Software, and are included by permission of Meade Instruments Corporation.

ISBN 0-306-45638-9 (Hardbound)
ISBN 0-306-45639-7 (Paperback)

© 1997 David H. Levy, Larry A. Lebofsky, and Nancy R. Lebofsky
Plenum Press is a Division of Plenum Publishing Corporation
233 Spring Street, New York, N.Y. 10013-1578
http://www.plenum.com

10 9 8 7 6 5 4 3 2 1

Printed in the United States of America

To the teachers and staff who participated in Project ARTIST. Their willingness to learn and their dedication to teach will increase the understanding of the universe and the appreciation of the sky for future generations.

Preface

A Sky Above, a Child Below

Picture a clear, warm evening, a quiet backyard, and a 10-year-old girl looking up at the stars. Does anyone live up there, she wonders. Who inhabits those majestic worlds? Probably creatures I can only dream of. How can I find out more about them?

As teachers and parents, we cannot give this child a complete response. No, Virginia, we do not know what fabulous creatures might live up there. But we can tell her, and thousands of other curious young children, about the night sky with its wondrous constellations and a multitude of suns and worlds.

Unlike the distant countries and lands we learn about in geography class, the sky is easy to find. Even from the middle of a big city, the brightest stars and planets and the Moon can be found on a clear evening; at least some sky should be available to virtually any child. The stories, history, and interesting sky-related facts could fill many a classroom hour.

For some reason, the sky has had a problem finding its way into school curricula. Even in this technological age of computers and video, where the latest pictures taken by the Hubble Space Telescope can be sent over the Internet directly into a classroom, most school districts do not encourage an activity as simple as looking at the sky.

The three of us who wrote this book come from different

A youngster's first taste of astronomy could be as simple as this look at the Moon. Photo by David Levy, courtesy Green Acres Day Camp.

backgrounds. David Levy has been an amateur astronomer for 35 years and brings a familiarity and sense of the night sky to his work with young children. He has discovered eight comets from his backyard and shared in the finds of 13 others with Gene and Carolyn Shoemaker. One of these comets, Shoemaker–Levy 9, struck Jupiter in 1994.

A professional astronomer, Larry Lebofsky studies asteroids, the solar system's smaller objects, found mostly between Mars and Jupiter. Nancy Lebofsky is a professional educator with experience organizing workshops for teachers. One day Nancy visited a classroom at her daughter's school and noticed a wall display showing phases of the Moon marked as half, full, and "gibbons."

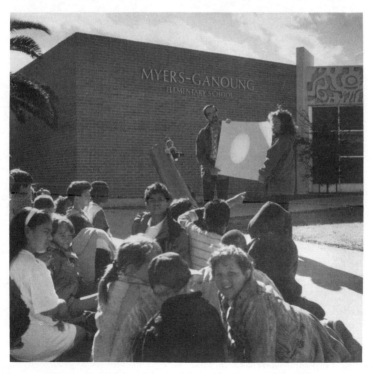

Larry and Nancy Lebofsky show students the Sun, safely! Photo by David Levy.

She and Larry discussed the implications of this error. Why was he studying remote asteroids, Larry wondered, when his daughter's school needed an explanation of what a gibbous Moon was? The Lebofskys decided to create a program through which teachers of young children could learn about astronomy and how to teach it.

We all share the basic idea that the sky should be presented simply and directly to as many young people as possible. An opportunity for this came after the 1994 comet impacts on Jupiter— an event that sparked so much interest that for a week almost

David Levy demonstrates the use of a telescope to a group of young children. Left to right: Teresa Fish, Doug Fish, David Levy, Andrew Roche, and Jerry Roche.

everyone was an amateur astronomer. Thousands were looking
through telescopes at a planet scarred with the darkest features
seen since the invention of the telescope. It was a golden opportunity to foster an awareness of the night sky. Now we should try
to build on that wonderful week and pass on to the next generation a sense of excitement about the sky and all that it offers. Then
we can follow the words of Horace, who hoped that with our
heads exalted, we would touch the stars.

> The sight of day and night, the months and returning years,
> the equinoxes and solstices, has caused the invention of num-
> bers, given us the notion of time, and made us inquire into the
> nature of the universe; thence we have derived philosophy,
> the greatest gift the gods have ever given or will give to
> mortals. This is what I call the greatest good our eyes give us.
> —Plato

Acknowledgments

Project ARTIST (Astronomy Related Teacher Inservice Training), our teacher enhancement program, was funded by the National Science Foundation, Grant #ESI-9154832. We thank the Lunar and Planetary Laboratory of the University of Arizona (UA), whose director, Dr. Michael Drake, has provided us with enthusiastic support. We also thank the UA's Steward Observatory, especially Donald McCarthy, and Larry Dunlap of the UA's Flandrau Science Center. Although many teachers, in and out of ARTIST, provided assistance, we specifically wish to mention that of Thea Cañizo, who provided many hours of help above and beyond the call of duty. The section "Alternative Activities" in chapter 6 is adapted from her article "Stellar Star Parties," *Science Scope* **18**, 6 (1994), 72–74.

David Levy also wishes to thank his wife Wendee Wallach-Levy, whose editorial assistance was much appreciated.

The following activities have been adapted from previously published ARTIST materials: Stained Glass Sun Symbols, Bear Hunt (*Science Scope Magazine*, 1994, 1997); Sun Symbols, Venus

Topography Box (*The Universe at Your Fingertips*, Project ASTRO, Astronomical Society of the Pacific, 1995); Magpies and the Milky Way, and Constellation Viewers (*Creative Childhood Experiences: Integrating Science and Math through Projects, Activities, and Centers*, 1997)

Contents

Part I

Opening the Sky

Chapter 1

Ideals Are Like Stars: An Introduction

Ideals are like stars; you will not succeed in touching them with your hands. But like the seafaring man on the desert of waters, you choose them as your guides, and following them, you will reach your destiny.
—SENATOR CARL SHURZ, in Faneuil Hall, Boston, 1861

We hear a great deal about the environment these days. But people of different ages and experience have very different ideas of what environment really means. A newborn child has its own pretty definite concept—centered on itself and extending to a room and a familiar face. As the child grows, its environment expands to take in a house and a street. With early education those surroundings quickly stretch to a country and a world.

At age 7, 17, or 70, that individual might look up and discover a sky filled with stars. That is when real understanding of "environment" can begin, an environment that includes the Universe. A teacher or a parent, by helping this process along, can steer it in some remarkable directions.

PURPOSE OF THIS BOOK

This book is about teachers and parents helping children to love the sky. This is not always an easy thing to do, for the sky does not appear accessible in the midst of a big city, nor is the night sky accessible during school hours.

How do we meet this challenge—to find a way past the cloud of difficulties to the starry sky? Each chapter in this book focuses on a different set of ideas. The chapter on observing the Sun, for example, explains how we can follow the life of one star in broad daylight. Another chapter shows how we can turn the Moon from a bright light in the sky to an actual place that children can even imagine visiting. Still another explores how we can turn the constellations from groups of stars into showcases of different cultures.

Although the book is directed primarily toward teachers and parents, it is not a workbook. More than anything else, we want our readers to begin to delight in the sky themselves, so that they will naturally share their enthusiasm with children. To that end, this book is far more than a collection of hands-on activities and demonstrations. It is designed to provide an understanding of what is happening in the sky so that our readers will be able to impart the excitement of discovery as well as knowledge to young people.

The rich lore of the constellations brings the sky down to a personal level, so we explore it early in the book. In chapter 3 we introduce the myths of the constellations from different cultures, from the Greek myth about Andromeda to the never-ending bear hunt, a beautiful Native American tale about hunters who pursue a bear around the sky each year. The children are encouraged to find these constellations for themselves and to invent their own interpretations as a special way to learn about the sky.

TELESCOPES FOR SKYGAZING

What is the best way to start out under a sky? There are almost as many answers to this question as there are amateur astronomers. Bob and Lisa Summerfield from *Astronomy to Go*, a Philadelphia-based education program, staunchly advocate the use of the biggest telescopes they can bring to the schoolyards.

1. *Teachers get enthusiastic at a National Science Teachers Association convention in Anaheim, California.*

They usually set up telescopes with mirrors that are 20 to 25 inches in diameter, with the idea that the children will not soon forget the excitement of looking through a very large telescope. In chapter 6 we will learn more about the energetic approach of *Astronomy to Go.*

On the other hand, coming into a schoolyard and declaring a holiday with a great big telescope might be fine, but for most teachers and children, after the holiday is over, all they are left with are great memories and a tiny, department-store purchased telescope. The letdown can be catastrophic. Why, they ask, do we fail to get any satisfaction at all out of this small telescope when it seemed so easy to use the large one? Worse, they sometimes conclude that it's their fault, that they just don't really know how to use a simple telescope.

Although most amateur astronomers strongly advise against buying cheap department-store telescopes (often called DSTs), the plain fact is that almost every child who owns a telescope owns a DST. We suggest that these telescopes be put to the best use possible; they should at least work on the Moon. Our program brings moderate-sized telescopes to schools, but we always also try to bring a typical small scope of the kind that most teachers and children will have to get accustomed to if they go out on their own. It is better, we feel, for the teachers and children to learn to use DSTs under expert guidance so that they will be comfortable with them. In chapter 5, we discuss this approach more rigorously.

KEEPING A RECORD

We believe that young observers will remember much more if they record what they see. The idea of keeping an observing log dates back to the earliest astronomers; there are surviving Babylonian and Chinese records more than 5,000 years old. Their modern descendants enjoy the sky from backyards, from schoolyards, and from summer camps. Modern observing records—even the ones kept by children—could be just as important as those from ancient times. In a real sense it doesn't matter what the observing log says or how it is put together. It can contain drawings, words, or anything that gives the reader an impression of the evening just past. A log can add pride to a young person's observation of the night sky, elevating him or her to the status of amateur scientist.

Keeping an observing log is a good habit to get into. Any child, teacher, or parent who keeps a log of observations will learn a lot more about the sky and get more out of a night under the stars. We emphasize not just writing but also drawing and other ways of keeping a notebook about the sky.

DAYTIME OBSERVING

School happens during the day, but it is a misconception that the sky "happens" only at night, which, as we have heard from some school districts, is a reason not to do astronomy. On the contrary, chapter 7 is about the most important star of all, the Sun. It suggests ways to observe the Sun safely and discusses how its magnetic storms, called sunspots, change their shape and structure each day as the Sun's rotation carries them across its surface.

Although it may be hard to visualize, the Moon is in the sky as much during daylight hours as it is at night. More effort is needed to see it during the day, but the majestic detail of its surface is still there. There is actually no better way to comprehend the orbit of the Moon around the Earth and the Moon's phases than to go outside each day and see where the Moon is in the sky.

THE SOLAR SYSTEM: A SENSE OF PLACE

The phases of the Moon are a means to get acquainted with the sense of the Moon as a body in space. This concept becomes important as we next try to introduce the other worlds of our solar system as places, all with their own environments, sunrises, and dark of night.

Children relate better to concepts and things that seem real to them. Chapter 8 looks at what it is like on the Moon and on each of the planets: how much a child would weigh, how hot or cold it would get, and what breathing the "air" would be like.

Our Venus Topography Box activity (see chapter 9), designed to show children how the Magellan spacecraft did its work as it orbited Venus, is a way of seeing their planet as a place. Forever shrouded in dense clouds, Venus' surface cannot be seen. But out of a shoebox, children create cloud-hidden environments on Venus and then figure out how to study these environments the way an orbiting spacecraft would. There is a way to "see" the surface through the techniques of radar. The idea is to give the children the impression that each planet is a special place.

COMETS AND IMPACTS

As the popularity of dinosaur movies so aptly demonstrates, children love big things. Dinosaurs are wild and fanciful, and so is the sky. With its many strange worlds, constellations, and vast distances, the sky offers a big area for a young mind to roam. Both topics are naturals for children, who can spend hours wandering around in the world of large and magical things.

Imagine the power of merging these two topics into a single unit on impacts. Some 65 million years ago, a large comet or asteroid slammed into the Earth and probably caused the dinosaurs to become extinct. The evidence for this is strong. In a thin layer all over the Earth is a high concentration of precious metals derived from an impacting body, as well as quartz that has been subjected to the heat and pressure of an impact. More important, we have found a "smoking gun," a large 65-million-year-old impact crater buried beneath the village of Chicxulub on Mexico's Yucatán Peninsula. Some 150 miles wide, the crater was formed by the impact of an object perhaps 10 miles across.

The subject of impacts brings together such seemingly diverse areas as our fragile biosphere, the ozone layer that protects it, how life evolved on Earth, and the kinds of objects that orbit the Sun. In chapters 10, 11, and 12 we discuss the topic first in general terms and then specifically the impact of Comet Shoemaker–Levy 9 into Jupiter.

EVERY STAR IS A SUN

It is a simple fact that our Sun is like the other stars in the sky and that every star is a sun of some sort. But this concept is difficult for children to grasp because it involves an understanding of the great distances in our galaxy. In chapters 13 and 14 we discuss the different types of stars, double stars, stellar evolution, and black holes, and suggest activities that help children to understand them.

As we move out beyond the stars, the whole Universe becomes a place with lots of galaxies and a place that is expanding. Most children see the Universe beyond the solar system as an abstraction, but in chapter 14 we will suggest ways to make this part of astronomy seem friendlier.

ABOUT ARTIST

The ARTIST program, which brought this book's authors together, was intended to open an avenue for teachers who want to add astronomy to the list of subjects taught in their classrooms. The name originally had nothing to do with art, being an acronym for *a*stronomy *r*elated *t*eacher *in*service *t*raining. However, once the Lebofskys had come up with the acronym, we realized that it would help bring home the interrelation of astronomy with other aspects of the school curriculum.

ARTIST's centerpiece was a 4-week-long workshop, held for the first time in June 1993 at the University of Arizona. (There was a shorter pilot workshop in 1990.) In addition to presentations by this book's authors, astronomers and teachers made presentations or led the group through hands-on activities. Each idea or project in this book should have a specific goal, should be adaptable to a range of age groups, and should be fun.

The elementary and middle school teachers were introduced to the basic concepts of both planetary sciences and astronomy, as well as to possible ways of making astronomy exciting in a classroom. On several clear nights, the teachers looked through a variety of telescopes, from binoculars and a simple refractor to Newtonian and Schmidt-Cassegrain instruments. In daytime sessions they learned how to examine the Sun safely and to measure changing shadows as the Sun progressed across the sky. Many of the ideas presented in this book were developed during the ARTIST workshops, in other workshops run by the authors, and by the teachers who attended these workshops.

A SENSE OF VASTNESS AND FRIENDSHIP

We want children to see the sky as something that is vast and friendly at the same time. The sense of vastness is there to excite and to prompt a child to look further. It is one thing to excite a child, but quite another to build on that excitement so that the sky remains a welcoming place. These goals are not mutually exclusive, and we kept them in mind when choosing the activities for this book.

The vastness of the heavens has been explained in several ways. The movie *Powers of Ten*, for example, leads us from a lakeside picnic dinner to the planets, the stars, and the great galaxy clusters. Our book tries to relate the environments on different planets to what a child might experience here on Earth. We also show how the Universe is big enough to offer many examples of exotic things, from an upside-down crater on the Moon to a star shaped like a teardrop to the ethereal beauty of the Crab Nebula.

That children should see viewing the sky as fun is the most important thing, for fun and interest lead to a long-lasting pursuit. Through the activities, stories, facts, and ideas presented here, we hope to make the sky a pleasant place both for children and for those who are fortunate enough to have the opportunity to teach them about it. What higher endeavor can there be than to open a child's eyes to the larger world, and the Universe?

Chapter 2

Some Basics: What Is Really Important?

The most important thing in starting a unit on astronomy is the spark of enthusiasm that gets the children wanting to learn more. We first should try to impress them with a sense of the vastness of astronomy's realm, with great distances, large sizes, and faraway places, as well as the idea that the solar system is, after all, our neighborhood. Here we will suggest some appropriate introductory strategies to open the world of astronomy to children.

The concept of place, introduced in chapter 1, is an important starter. The Moon is an object that can be observed from anywhere, with or without a telescope, and it's easy to see that it has a surface with different types of terrain. The dark and bright areas, which have been variously referred to as the Man in the Moon or the Rabbit in the Moon, are among the most obvious features that anyone can see in the sky. We should take advantage of this, because a simple look at this shading tells us something about the Moon as a world. The bright areas are mountain ranges and highlands, and the dark parts are lava plains.

LEARNING THE SKY

Feeling comfortable with your subject before you begin teaching it is an important step; for some this is a major hurdle, as there is precious little material available to most teachers and parents to help in organizing an astronomy unit. It is also important to remember that working on the unit should be an enjoyable experience both for you and for the children.

The more confident you feel about knowing the sky, the more effective your unit will be. So before you embark, try to become familiar with the sky on your own. Keep an observing log as you begin to follow the simple motions in the sky, such as the position and phase of the Moon as it moves from night to night.

By using a star at a time as a guide, you can "star hop" your way to becoming familiar with the sky. For example, the two stars at the end of the Big Dipper's bowl point toward Polaris, the pole star (see Figure 2). Just find the Big Dipper, mentally draw a line joining the two stars at the end of the bowl, extend it five times, and you're at Polaris.

All this is easy *if* the Big Dipper is in the sky, which from midnorthern latitudes is every evening of the year. However, in fall and winter you may not see it unless you have a good northern horizon. From the southern United States, the Dipper is below the horizon during late fall and winter evenings.

The Big Dipper is a key that can lead you to other stars and constellations in the sky. Joining the three stars in the Dipper's handle forms a curved line or arc. Extend the line away from the Dipper to "arc to Arcturus," the brightest star in Boötes, the herdsman, and then "speed on to Spica," Virgo's brightest star, in the same direction (Figure 3).

If the Dipper is low, then Cassiopeia, on the other side of the pole, is high. It does not point the way as obviously as the Dipper, but it will give you a general idea of the direction of the pole star (Figure 4). Finding the pole is important because that is the basis of orienting yourself to the sky. The Earth rotates in a day around an axis that points north and south. In the northern sky, that axis

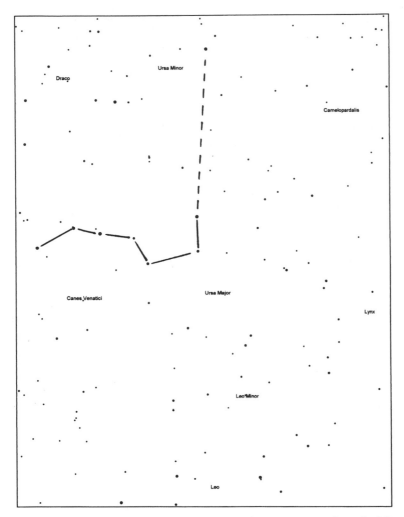

2. *The Dipper's pointer stars and Polaris. The two stars at the end of the Big Dipper point to Polaris, the North Star.*

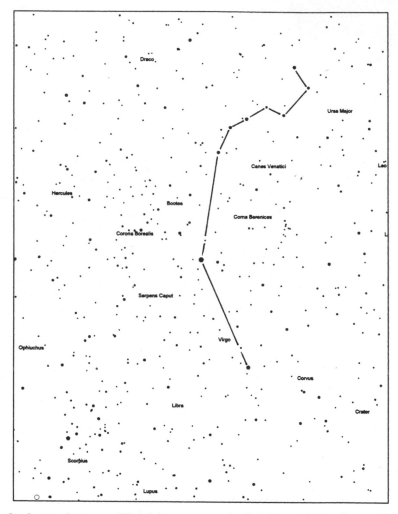

3. *Arc to Arcturus. The three stars in the Big Dipper's handle can be joined with a curved line or arc. Extend that line to Arcturus, in Bootes, and continue to Spica, in Virgo.*

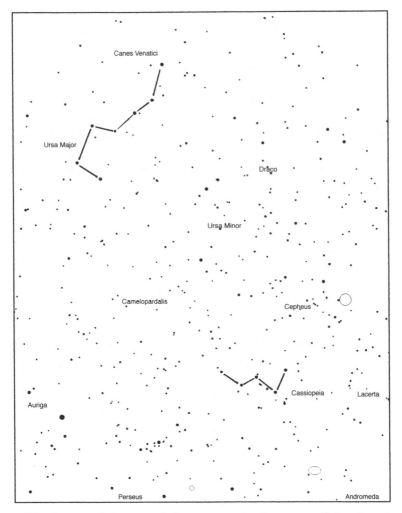

4. *The Dipper, Polaris, and Cassiopeia. At the center of the chart is Polaris, the polestar (or North Star).*

points roughly toward Polaris, and in the southern sky it points toward the faint star Sigma Octantis (Figure 5).

The sky of the Northern Hemisphere winter has so many bright stars that it is easy to find your way by going from one bright star to another. We recommended the route of "the heavenly G" (Figure 6), which is particularly useful since it goes through several constellations at once. Begin with Aldebaran, the bright red star in Taurus, and continue through Capella in Auriga and the brightest stars of Gemini, Castor, and Pollux. The underpinnings of the G are Procyon, the brightest star of Canis Minor, and Canis Major's Sirius, the brightest star in the sky. The G's curve stops in Orion at Rigel and indents from Rigel to Betelgeuse.

These are just examples to get you started. You can devise your own star-hopping journeys. Just move along the sky a star at a time until you reach your destination. The important thing is to take your time and get to know the sky, star by star, constellation by constellation, on your own terms.

Spend as much time as necessary getting familiar with the night sky before you begin your unit or try to teach your own children. It is not hard. If you use the step-by-step approach we have outlined, it should be easy and fun. At the same time, you will notice how the sky changes slightly as the nights go by. This is not obvious at first, for each star rises about 4 minutes earlier each night. But that adds up quickly, as we shall discuss later, to ½ hour a week, or 2 hours a month.

Now, what about the planets? Venus, Mars, Jupiter, and Saturn are bright enough that they may be confusing at first, because the fact of their changing positions means that they cannot be plotted on star charts. But you can identify them as planets. They will all be in zodiac constellations and follow the ecliptic, which is marked on a star atlas. Venus, the brightest, is never very far from the Sun, so if it is in the evening sky it will always be in the west and in the morning sky it will be in the east. Jupiter is much brighter than the brightest star, Saturn is as bright as a typical bright star, and Mars varies considerably in brightness depending on its distance from Earth. However, if you look carefully you may be able to notice its reddish tinge.

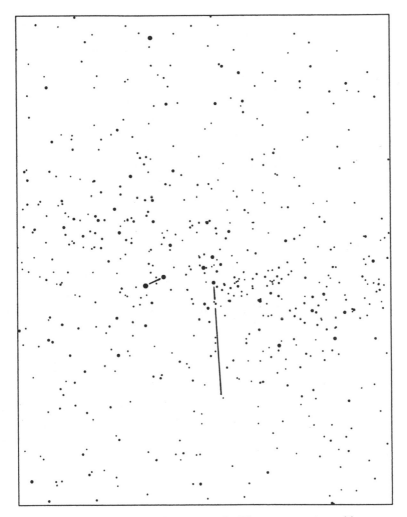

5. *The Southern Cross and the south pole. The cross consists of four stars roughly outlining the shape of a cross. The line points from the cross's southernmost star to the pole. The two stars left of the cross are Beta and Alpha Centauri; Alpha is the closest star to the Sun.*

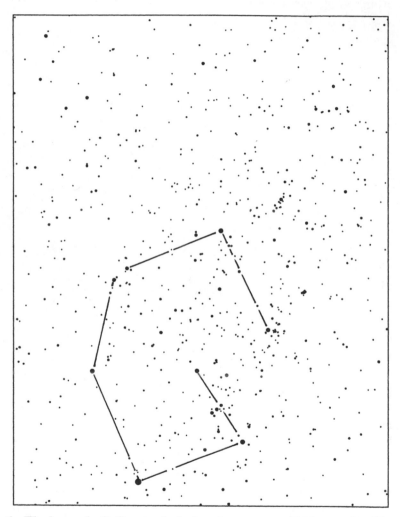

6. *The heavenly G. Counterclockwise, the stars are Aldebaran, Capella, Castor, Pollux, Procyon, Sirius, Rigel, Orion's belt, and Betelgeuse.*

If you are in an urban setting, finding even the major constellations may be hard because there are simply not enough stars visible. However, before you decide that this is the case, give your eyes a chance to adapt to the darkness. It takes at least 10 minutes for your pupils to dilate, a process that increases their sensitivity to the night sky. If you give your eyes this time, you will see that more stars are visible and the outlines of some constellations become clear.

Where the sky is very bright or hazy, though, you may never see enough stars to make out some basic patterns, and you may have to invent your own: Someone we know made up her own geometric figures, such as triangles, squares, and arcs, based on the few stars she could see.

If you are in a suburban location, your night sky should allow you to identify the major constellations easily. Avoid observing near bright lights that come from windows or from street lamps. If you are fortunate enough to live far from any city lights, your night sky could be spectacular, especially if there is no bright Moon. However, the downside to a dark sky is that it can be just as much a hindrance to finding constellations as a bright one. Under such a sky, you also need a longer period of time—at least 20 minutes—to adapt fully to the darkness. Once you are adapted, you will be surprised at how much there is to see. You can see as many as 3,000 stars—far too many for it to be easy to identify patterns. If you find this to be a problem, try becoming familiar with the constellations when the Moon is in the sky, as its brightness reduces the number of stars you can see to a manageable level.

ASPECTS OF THE SKY

DAY AND NIGHT.   We can visualize two things about the Earth's motion just by going outside and looking up. We know that the Earth turns on its axis once in a little under 24 hours, because we can see the stars rise and set in this amount of time. We perceive as well that the Earth also revolves around the Sun. As the Earth moves, our base of observations changes, resulting, as we mentioned earlier, in the stars rising and setting about 4 minutes earlier each night, or 2 hours earlier each month.

To aid in understanding aspects of the sky, it helps to imagine that the sky has a number of reference points and lines. Although we never see these lines, knowing what they mean will help us understand why the sky changes from hour to hour, and from night to night, and why we see what we see up there. We call the lines *great circles*, because they are the largest possible circles that can be drawn on a sphere—in this case the celestial sphere.

Zenith.   The first invisible point you should know about after you go outside is the point directly over your head. That point is called the zenith.

Meridian.   This is an imaginary great circle passing through your zenith and the celestial pole near Polaris. This line is important because when an object crosses it, that object is as high in the sky as it's going to get. The Sun crosses this line at about noon every day, which is why the morning hours are called AM, for ante meridiem, and the afternoon hours are called PM, for post meridiem. We say that the Sun, or any star, culminates when it crosses the meridian.

Altazimuth Coordinates.   With zenith and meridian in mind, it's easy to think of the sky in altazimuth, or altitude and azimuth, coordinates. The altitude, or elevation, divides the angular distance from horizon to zenith into 90 intervals called degrees. The azimuth goes round the horizon from north, which is 0 degrees, to east (90 degrees), to south (180 degrees) and west (270 degrees).

Every hour of every day, a panorama unfolds right over your head, and all because the Earth is spinning eastward. The consequence of this simple fact is that we get day and night.

The Ecliptic.   The ecliptic is the apparent path of the Sun around the sky in the course of a year. The solar system is on a flat disk.

Since the Earth is in the plane of the solar system, the other planets tend to stay fairly close to the ecliptic in the sky. For this reason, ancient people attached great importance to constellations on the ecliptic, a group known as the *zodiac*. A nondescript band of constellations in the sky, the zodiac is significant in astronomy because the planets, Sun, and Moon can usually be found within it.

WHAT APPROACH TO ASTRONOMY SHOULD I USE?

Depending on your own interests, the children you are working with, and the sky conditions in your area, you have a choice of approaches. You can base your entire unit, for example, on the myths and legends that different cultures have developed. Your outdoor sessions would then focus on pointing out the star patterns as different cultures saw them (see chapter 3). Or if you want to emphasize the solar system, your observing sessions can highlight watching how the planets move through the sky over several months, an activity that does not require a telescope. Later you can try binoculars, and later still you can see how the Moon and planets appear through the eyepiece of a small telescope. Another approach would emphasize the constellations as part of the vastness of our Universe of stars and galaxies.

WORKING WITH NATIONAL STANDARDS AND STATE AND LOCAL FRAMEWORKS

In recent years, several organizations have published "national standards" for science education from kindergarten to twelfth grade. These include among others the American Association for the Advancement of Science's (AAAS) *Science for All Americans* and the National Research Council's (NRC) *National Science Education Standards*. These recommendations are now being implemented by individual states in various frameworks. Some of these frameworks are very specific, while others are more general, giving local districts flexibility in how the standards are to be achieved and thus what can be taught at any grade level in any science subject.

Therefore, what teachers will be able to teach in astronomy and how they will teach it will be very much dependent on state and local policy. However, our experience has shown that interested teachers can be very inventive and have been able to integrate astronomy into whatever they are required to teach at their grade level.

WHAT TO EMPHASIZE

Why not begin with the solar system? The Sun and the planets are worlds with distances that while large are measurable in kilometers or miles rather than light-years. These worlds are places and some of them have rocky surfaces. It is even conceivable that one of your children could someday walk on the surface of Mars.

The important thing is that children get some idea of what it would be like to live on these worlds. Relating familiar concepts to these unfamiliar places might help. What would the weather be like, for example, on Mars? on Pluto? How much would a child weigh on a planet as massive as Jupiter or on a small asteroid? Answers to questions such as these give the children a feel for what life might be like in these hostile environments.

Presenting the constellations is a different matter. These figures in the sky are our own creations and reflect the wisdom of cultures that have come and (for some) gone. There is as well a more important theme here. Fantasy and legend allow the mind to roam. A young child familiar with the traditional stories can spend many happy hours looking at the stars, connecting them in original ways, and creating new tales about wild and fanciful characters. This is the best way to learn one's way about the sky. With a sense of adventure and imagination, a child can explore the sky and make it personal, inventing private constellations and making up stories that fit.

Most astronomy programs spend too little time discussing stars, which for many people is what astronomy is all about. The reason for this is that stellar astrophysics is a complex subject, and many education texts do not give it much attention. But stars are fun to study. They come in all sizes and several colors, many are paired up as double stars, and some change in brightness. Some stars end their lives in the blaze and glory of a supernova, when for a few months they will outshine the combined light of all the other stars in the galaxy. More than light is involved in such an explosion, though. In its dying moments, a star that explodes as a supernova is creating the heavy elements, such as carbon, that are the bases for life.

After the supernova stage, what remains of the star's core begins to contract, sometimes into something called a neutron star or a pulsar and sometimes, if the core is massive enough, into a black hole. Children are excited by the idea of black holes, even though neither they nor their teachers, indeed not even many astronomers, really understand what black holes are.

When we explore galaxies and the expansion of the Universe, we run into even more problems. A galaxy is a collection of stars so huge that our minds are unable to comprehend the dimensions. Our own Milky Way stretches for more than *100 thousand* light-years—that means that traveling at the speed of light it would take us 100 thousand years to get from one end of the galaxy to the other, which is more time than any science fiction show allows for such a voyage.

WHAT TO AVOID, AND THE DANGER OF OVERSIMPLIFICATION

Rote Learning

It is difficult for a child to have to memorize a series of seemingly unrelated facts. Although it is hard to avoid this entirely, we can make it fun. To learn the names of the planets in the order of their distance from the Sun, for example, we try the mnemonic "*my* *v*ery *e*ducated *m*other *j*ust *s*erved *u*s *n*ine *p*izza pies" to the tune of Stephen Foster's "Old Folks at Home."

Mnemonics are more useful, of course, if they actually make sense. "*Oh be a fine girl* (or *guy*), *kiss me*" is the old sentence by which we remember the colors, or spectral types, of the stars—O, B, A, F, G, K, and M. We can add "*right now*" for the rare classes R and N at the red end of the spectrum. The jingle doesn't teach us anything about what a spectrum is or why spectral type is important, but astronomy students have been memorizing it for years.

What is much more important than the mnemonic is the fact that a simple look at the sky on a clear night can tell us the spectral type of several bright stars. In the northeast corner of Orion, for example, is a bright reddish star called Betelgeuse, and at the opposite, southwest corner is a bright bluish star named Rigel. Rigel is in spectral class B, a grouping reserved mostly for blue giant stars. Betelgeuse, on the other hand, is an M class red giant star. Many of the sky's brightest stars show these colors—not obvious colors, like red lights, but subtle hues that you should be able to distinguish with a little practice.

Difficult Subjects

Much as we try to make the sky simple and inviting, we often are confronted with the fact that at some level astronomy is a complex science involving math and physics. We can explain the basic concepts as simply as possible, but if we try to oversimplify certain topics, we can fall into inaccuracy and error. Relativity and black holes are two significant examples of such topics. We should not ignore them. But how do we discuss relativity—a subject that Einstein introduced to the world less than a century ago, which many professional astronomers still don't fully understand?

Our approach is to tread lightly. With relativity we try thought experiments, the same tool that Einstein himself used. We have the children imagine taking relativistic trips through space as their friends stay behind and then see what happens when the young travelers return to aged friends. The beauty of this strategy is that it allows the children to take a look at a complex subject without being overwhelmed by it.

In the same vein, the concept of black holes, where a dying star's gravity is so strong that not even its own light can escape, is so naturally appealing to children that it would be a pity to ignore it just because the physics involved is so complicated. However, some attempts to make black holes easy—a movie called *The Black Hole* from around 1980 comes to mind—are so inaccurate that they leave their viewers with a short-sighted idea of what black holes really are.

Ultimately, teachers of astronomy try to steer a middle course between the extremes of oversimplification, on the one hand, and ignoring a subject altogether because it is so complex on the other. Our philosophy in this book is to introduce these subjects at the risk of erring on the side of simplification. However, we try to ensure that our explanations and suggested activities do not over-simplify to the point of being misleading.

TAKING ADVANTAGE OF SPECIAL EVENTS

Teachers have many motivational tricks that are designed to get their students excited about the world around them. But in astronomy, the sky itself adds to the fun by providing its own stunts, which are natural attention-getters. Eclipses of the Sun and Moon are the most obvious attractions, and they are a source of excitement and uncertainty. The excitement lies in the fact that something unusual is about to occur; there is uncertainty because most people have never seen an eclipse before and don't know what to expect, AND viewing the eclipse at all depends on weather prospects, which are often uncertain right up to the moment the show begins. Also, the same type of event might not occur again for a long time.

Alignments of planets as they move through the sky can also be exhilarating to watch over a season. When two or more planets seem to approach each other in the sky, or when a planet appears to pass by a star, we can watch a celestial dance that goes on night after night.

Man-made events can be exciting, too. The space shuttle's frequent flights often take it over areas where it can be viewed by large numbers of people, either just after sunset or just before sunrise. Since the shuttle, as all satellites, shines by reflected sunlight, it is invisible during the darkest hours of night when it too is in darkness. A rare exception to this occurs when the shuttle is about to make a night landing at Florida's Kennedy Space Center. As it descends it is sometimes visible across the United States as it paints a bright orange arc across the sky.

Finally, since the early 1970s, amateur astronomers, museums, and planetariums have celebrated Astronomy Day, a springtime festival that is usually held on the Saturday nearest the first quarter Moon in April or May. There are shopping mall displays and observing sessions under the stars.

A FINAL THOUGHT

Some years ago a family walked into a planetarium theater in Vancouver, Canada, for the scheduled presentation. Father, mother, and an autistic son who had hardly said a word in years sat down. The lights went out, and the show proceeded with stars, special effects, lots of music, and narration. The show ended with a magnificent sunrise. As the lights came on again the child turned toward his parents, opened his mouth, and said "star." The parents were ecstatic. When the planetarium staff heard this news they immediately took the child on a special behind-the-scenes tour and then set up a program for him that was designed to make him feel comfortable in the planetarium environment. A single word will not heal a lifetime of illness, but it can be the first vital step of a long journey.

Given enough encouragement and the right kind of teaching, many children can be made to feel comfortable among the stars.

Chapter 3

Cultures and Constellations

Over thousands of years people of many civilizations, influenced by climate and history, have tried to explain patterns of stars in terms of their own cultures. These patterns or constellations include such creatures as a bear, a hunter, a dragon and serpents, and a giraffe. Although some bright stars have been given several different names, most of the commonly used ones derive from the early Arabs and Greeks.

Trying to relate the star patterns to their lives was a prime goal of early skywatchers. It is impossible to say who created the first constellation. We have records dating at least as far back as the ancient Chinese of 1,600 years ago. They had 28 sections called *hsiu*. In India, a system of *nakshatra* dates back almost 1,000 years. The Chaldeans developed some of the constellations we recognize as far back as 5,000 years ago. Around 150 A.D., Ptolemy completed his *Almagest*, which included a listing of more than 1,000 stars arranged into 48 constellations.

By the start of the seventeenth century, the German astrono-

7. *Multicultural skylore takes place in a classroom.*

mer Johann Bayer was ready to set the known stars and star patterns down in an organized way. His *Uranometria*, published in 1603, listed stars by Greek letters, a system in which the brightest star was generally called Alpha (such as Alpha Orionis), the next brightest Beta, and so on. Early in the eighteenth century, John Flamsteed, an English astronomer, extended the list of stars by adding Arabic numbers, such as 2 Lyncis, to fainter stars.

A RICH TAPESTRY OF STAR NAMES

Coming from virtually the dawn of history, the names of the stars above us are strange and fun. One favorite is Zubenelgenubi, the ancient Arabic name given to Alpha Librae, the brightest star in Libra. Both the Greeks and the Arabs saw this star as representing the southern claw of nearby Scorpius, which is what this splendid name means. Beta Librae, the old northern claw, is called Zubeneschamali.

The fainter the stars get, the less interesting their names, such as Flamsteed's 2 Librae, best seen through binoculars, and the Smithsonian Astrophysical Observatory's 158846, a star too faint to be seen without a telescope. One star of recent interest is 51 Pegasi, a star that might have a Jupiter-sized planet orbiting in its outer atmosphere as well as other planets. Racing around the star in a little more than 4 days, this planet must have a hard time avoiding destruction from the tidal force of its nearby sun.

Betelgeuse is another strange name. Some older texts incorrectly advised us to pronounce the name of this star as "Beetle Juice," although we wonder why a star of such majestic brilliance would be diminished to liquid refreshments for bugs. *Bet*-el-jooz is a more appropriate pronunciation for this name, which refers to the shoulder of Orion. Mintaka, the delightful Arabic name attached to the westernmost star of Orion's belt, means simply the belt (Figure 8).

STARS IN SCIENCE AND MYTH

When we look at the sky each night, we see a parade of stars that represent different colors, types, sizes, and ages. But thanks to the richness of our cultures, we also see them as forming shapes that had meaning to different civilizations. For many children, this part of the stellar adventure is the most entertaining.

The best way to introduce children to the constellations is not to. Their first night out under the stars should offer a sense of adventure, where each young mind will come up with new ideas and different shapes. The children's first nights should be nights of discovery, a chance to set out on their own. If ancient peoples all over the world, such as the Egyptians and the Greeks, had the chance to create a mythology about the star patterns, why shouldn't modern children?

In the fall sky, which is the one most easily available to teachers beginning a new school year, the easiest pattern of stars to begin with is the Square of Pegasus. Although this pattern is not easy to give directions to, it is actually fairly conspicuous in the

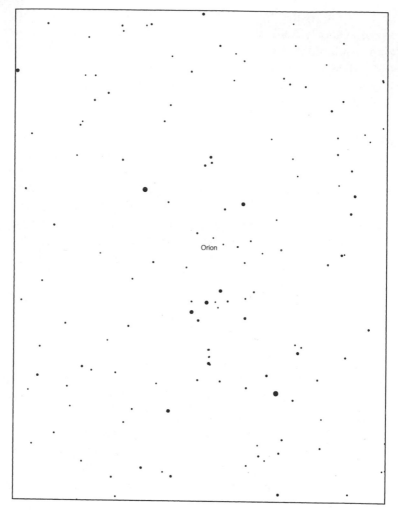

8. *Orion. Betelgeuse is the bright star at upper left; Rigel is at lower right.*

sky. Consisting of a rectangle of four moderately bright stars, the square occupies an uncrowded region.

The traditional story about Pegasus also includes the neighboring constellations of Cassiopeia and Cepheus, the queen and king of Ethiopia, and their daughter Andromeda. Boasting that

Andromeda had incomparable beauty, Cassiopeia angered the god of the sea, who chained Andromeda to a rock and sent the monstrous whale Cetus after her. Perseus came to the rescue just in time, taking the princess, her chains, and the rock on the winged horse Pegasus. All the figures in this story are represented in the sky at this time of year, adding life and intrigue to evening skywatching sessions. The story and view also inspired John Keats, who wrote these words about the princess in "Endymion":

> Andromeda! Sweet woman! why delaying
> So timidly among the stars: come hither!
> Join this bright throng, and nimbly follow whither
>   They all are going.

Although the constellation story is an intriguing one, we think that its greatest value is in opening the eyes of children to possibilities of different forms in the sky. In modern America, for example, could the square not represent a baseball diamond, that icon of present day North America, which is much more relevant to children than a princess and a horse? The diamond has a triangle of stars at home plate, where one represents the base while the other two are the batter and catcher. Second base is the star Alpheratz; although it is technically within the boundaries of Andromeda, it is also an integral part of the square. There is also a star at third base. A faint star within the square is the pitcher. Finally, the nearby stretch of the Milky Way represents the cheering fans in the stadium (Figure 9).

ACTIVITY: CREATING A BRAND NEW "CONSTELLATION"

PURPOSE: To open young minds to the possibility of inventing their own patterns in the sky.

MATERIALS: Flashlight with strong beam
    A clear night: passing clouds make constellation watching more difficult

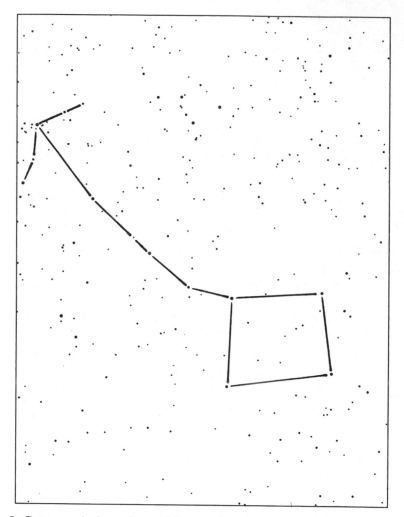

9. *Pegasus, Andromeda, and Poseus. The four stars of the square of Pegasus join at left to the thin line of Andromeda. Perseus is at upper left, and Cassiopeia is near the top.*

PROCEDURE:

1. Gather the children on the side of you opposite where Pegasus is (Figure 9).
2. Give them a chance to look at the placement of stars in the sky.
3. Point your light first toward Cassiopeia. Ensure that the children can see where it is pointed. Now point it toward Cepheus, then Pegasus, Andromeda, and Perseus. Tell them the popular mythological story as you go along.
4. Suggest the baseball diamond as a possible modern interpretation. Tell them how the diamond works in the sky.
5. Give the children time to look again at the stars, and suggest that they invent their own constellations, using their own experiences, family members, friends, pets, or political figures to place in the sky.
6. Once inside, ask them to write about or draw the new constellations they have invented.

CLOSURE: Discuss what new figures and images the children have come up with and encourage them to expand their private sky maps with each clear night.

SOME MAJOR CONSTELLATIONS

Following is a selection of major constellations, each coupled with a description of its mythological significance. Each constellation name is followed by a simple description of what it is and a phonetic key to pronunciation. We cover the constellations that are visible all the time from most of the United States and Canada, then explore the constellations of the four seasons, and finally look at a few forms that are visible from the Southern Hemisphere. Within each section, constellations are discussed alphabetically.

North Circumpolar Constellations (Figure 10)

Not all stars rise and set. From North America, far northern stars appear to circle the pole, staying in the sky all night long, all year round. Here are a few:

Cassiopeia—Queen of Ethiopia (cass-ee-yo-*pee*-a)

One of the most easily found constellations in the sky, Cassiopeia looks like a W or an M consisting mostly of a triangle of three stars with two fainter stars nearby. As we have already seen, the Romans saw Cassiopeia as the Queen of Ethiopia chained to her chair, condemned to circle the sky throughout eternity, sometimes right side up, and sometimes upside down, as a punishment for boasting of her daughter Andromeda's beauty. Arabic cultures saw the same constellation as a kneeling camel.

Cepheus—King of Ethiopia (*see*-fee-us)

Considering its mythological fame, Cepheus is an inconspicuous constellation, its five bright stars easy to find only because they face the open side of Cassiopeia's W shape. Cepheus was known to the inhabitants of the Tigris-Euphrates Basin as early as 2000 B.C. One of Jason's Argonauts, he was Cassiopeia's husband and Andromeda's father. The myth of Argonauts appears all over the northern and southern sky, from Cepheus and Hercules in the north to the parts of their ship *Argo* in the south. Cepheus looks a little like a house with a very pointed roof. Although the top of the roof does not really point to Polaris, it does offer the general direction to the pole at a time of year when the pointer stars of the Big Dipper are not readily accessible. Cepheus works as a pointer because there are no other bright stars near Polaris.

Draco—The Dragon (*dray*-ko)

This constellation is circumpolar from all but the extreme southern parts of the Northern Hemisphere. At its highest, at midnight in

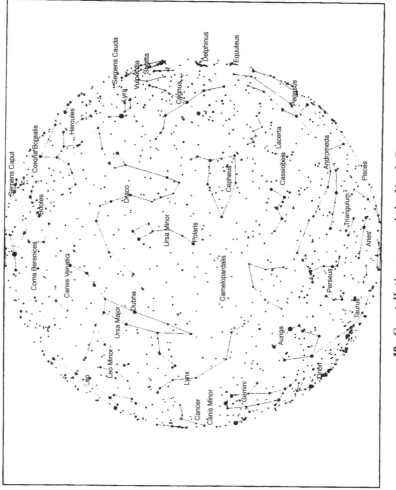

10. Constellations around the North Star.

April and May, Draco is best seen during the Northern Hemisphere's warm months. A large and faint constellation, the Dragon is hard to trace as it winds about other constellations, especially Ursa Major. The Chaldeans, Greeks, and Romans all saw the figure as a dragon, while Hindu observers saw an alligator. The Persians saw a man-eating serpent. The proximity of Draco to Hercules works well with a classical story: Hercules killed Draco as the dragon defended the golden apples in the garden of the Hersperides. Thuban, the brightest star in the constellation, was the pole star that the Egyptian pyramids pointed to at the time of their construction (see chapter 8 for a discussion of precession).

Ursa Major—The Great Bear (*ur*-sah major)

Ursa Major is the subject of many rich legends and interpretations. A taste of the panorama of human imagination:

Although most North Americans see part of the bear as the Big Dipper, many Native Americans do not. The Sioux of central North America see a long-tailed skunk, and the Eskimos see a kayak stool and reindeer. The ancient Aztecs saw Tezcatlipoca, an evil god responsible for many conflicts, dancing forever around Polaris.

In the Cherokee legend, the handle represents a team of hunters pursuing the bear.

Two northern tribes, the Iroquois of the St. Lawrence River Valley and the Micmacs of Nova Scotia, have embroidered on the story. The bear in their legend consists of just the Dipper bowl, and seven warriors are hunting it. Each spring the hunt begins when the bear leaves her den in the semicircular constellation we call Corona Borealis, the Northern Crown. In autumn's hunting season, the bear is finally killed, but only after some of the southernmost hunter stars have given up. Then the skeleton, now lying on its back, remains in the sky until the following spring. Meanwhile a new live bear is hibernating in Corona Borealis, ready to emerge in the spring, and a new hunt starts.

In a Chinese legend, the stars form a bushel to deliver food in fair amounts to the population in times of famine, a role also assigned to Sagittarius. Some 5,000 years ago the Chinese had a unique legend: The region represented the Jade Palace, where Wen-chang, the god of literature, gave advice to K'uei, Minister of Literary Affairs. Joining the discussion are Chu-I, Mr. Red Coat, the minister who looks after student welfare; Chin-choa, Mr. Gold Armor, a minister who searches for talented young people; and Kuna-ti, a god who prevents war. No doubt this interpretation comes from ancient generations of Chinese university students. Ancient Hebrews also saw a bushel, and an Egyptian myth pictures the Dipper stars as the hind leg of the bull.

The British see a plow, although early Britons saw King Arthur's chariot; an old Welsh legend also equates the stars with King Arthur. On clear nights the sky over Germany has a great wagon and three horses. The Romans viewed a team of seven oxen, tied to the pole and driven by Arcturus. A North African scene involves a camel, and in the East Indies they picture a shark, a hog's jaw, or even a canoe. Other cultures have seen the stars as seven sages or seven wise men.

Ursa Minor—The Little Bear (*ur*-sah minor)

Considering its importance as the constellation of the north celestial pole, Ursa Minor is actually a rather faint grouping. Also known as the Little Dipper, it looks more like a spoon whose handle has been bent back by some playful child. Some 2,600 years ago the Greek astronomer Thales first recorded it as a bear. Later, Phoenicians called it a guider. Observers in Denmark called it a small chariot.

Constellations of Northern Hemisphere Spring (Figure 11)

Boötes—The Herdsman or Plowman (boh-*oh*-tes)

Forming the constellation of Bootes, Arcturus and its neighboring stars make up one of the easiest patterns to find. Join the three

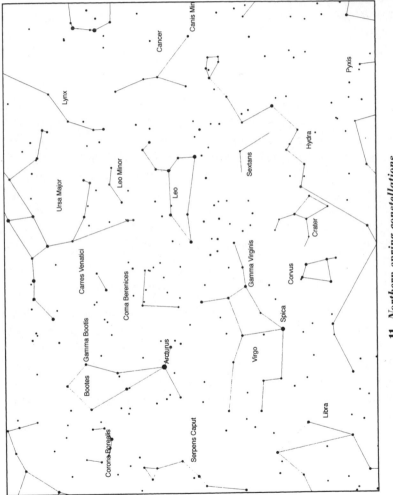

11. *Northern spring constellations.*

stars in the handle of the Big Dipper, arc to Arcturus, and you're there. The brighter stars outline a kite flying about. If you see the Big Dipper as a plow, as English tradition does, then Bootes is Arcas, son of Jupiter and Callisto, who drives the plow around the pole. Or he is a hunter like Orion, whose dog is nearby Canes Venatici.

Cancer—The Crab (*can*-sir)

This faint constellation occupies the empty space between two of the sky's most easily seen constellations, Gemini and Leo. Most clearly seen in January, it contains no bright star. Were it not for its role as a member of the zodiac and for its beautiful Beehive star cluster, we would have little to write about it. The Chaldeans and Platonists thought of Cancer as the "Gate of Men," which was opened for souls to gravitate to Earth to animate newborn babies.

Coma Berenices—Berenice's Hair (*ko*-ma bear-eh-*nice*-ease)

Like Cancer, this constellation, which lies nestled between Arcturus and Denebola (Beta Leonis), has no bright stars. It is essentially a cloud of faint stars, and through binoculars you should see at least 30 of them. Berenice, the beautiful wife of the Egyptian king Ptolemy Evergetes, sacrificed her long hair to Venus when her warring husband returned. The hair was placed in the temple, but someone stole it. A furious Evergetes was about to put the temple guards to death when he was told that Venus herself had taken the hair and put it in the sky not far from Arcturus. The king believed the story and spared the guards.

Corona Borealis—The Northern Crown (*ko*-row-nah bor-ee-*ah*-lis)

Just 20 degrees northeast of Arcturus is the semicircle of stars that form the Northern Crown, and although this constellation is small and faint, it is very distinct.

In the most popular of several stories, this crown belongs to Ariadne, daughter of the king of Crete. Ariadne refused a marriage proposal from Bacchus, as she did not wish to marry a mortal. To prove he was an immortal god, Bacchus yanked off his crown and threw it up into the heavens, as he had done for Corona Australis. Finally satisfied by this considerable sacrifice, Ariadne married him.

As we have seen, for some North American cultures Corona Borealis represents the cave where the bear hibernates each winter.

Corvus—The Crow (core-vus)

This small foursome of stars that the ancient Hebrews, Greeks, and Romans all called the Crow or Raven is easy to find in the sky of winter and spring just west of the bright star Spica. The crow's sad tale began simply enough, when Apollo sent him for a cup of water. After delaying for a long time, Corvus returned with the cup of water and a big water snake. He explained that the snake had attacked him. Knowing that Corvus was lying, Apollo placed him in the sky atop the serpent. Further, he placed the cup of water, now called Crater, nearby on the serpent in such a way that Corvus could never drink from it.

Hydra—The Female Sea Serpent (high-drah)

Trying to find Hydra, the sea serpent, in the southern sky, might be as difficult a task as Hercules had when he tried to kill it. The serpent that Hercules encountered had nine heads, and each time he lopped off one, two others grew in its place. Ever resourceful, Hercules set each severed neck afire, so that new heads could not grow. To add spice to the battle, Juno sent along a crab, which Hercules stepped on. Hercules is in the sky, and so are the crab and the snake, but at a safe distance. Cancer, the crab, lies north of Hydra's head.

Leo—The Lion (lee-o)

Leo is highest in the sky in March: one wonders if "March coming in like a lion" refers to this. Unlike most of the constellations, this figure, with its sickle tracing out a great head, really does look like a lion. Several cultures, especially the ancient Babylonians, suggested that Leo was connected with the Sun, because the summer solstice occurred when the Sun was in Leo. The Egyptian Sphinx might evoke the celestial Leo, as does the English royal standard.

Libra—The Scales (*lee*-bra)

Among the ancient observers, the Greeks saw this area of sky near Virgo as a part of Scorpius and did not put a balance there. The Chinese, Egyptians, and Hebrews all adopted a figure holding scales, but it was the Romans who attached the name Libra to the constellation.

Serpens—The Serpent (*sir*-penz)

Serpens is the animal that Ophiuchus holds as he goes about his business of healing the sick. It is the only constellation that is divided into two parts, Serpens Caput and Serpens Cauda (head and tail). Ophiuchus separates the two parts. The region is best seen during May and June.

Virgo—The Maiden (*vur*-go)

Culminating at midnight early in April, Virgo is easy to find by using the "arc to Arcturus, speed to Spica" routine. Spica is Virgo's brightest star. A very old constellation, Virgo was the maiden of ancient India. In a beautiful Greek myth, Virgo was the goddess of justice, Astraea, who with other gods walked with humans during the Golden Age. But as society began to disintegrate, the gods abandoned the land for the security of Heaven. The most patient of the deities, Astraea, was the last to depart, as Virgo, sadly carrying her scales of justice, which still ride with her across the sky.

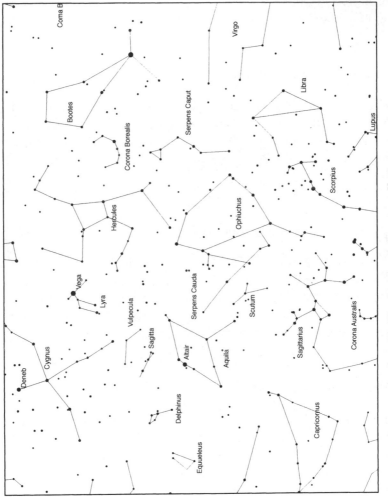

12. *Northern summer constellations.*

Constellations of Northern Hemisphere Summer (Figure 12)

Aquila—The Eagle (*ah*-kwi-lah)

Easy to find because of its bright star Altair, Aquila belonged to the Greek god Zeus, and his main accomplishment was to bring the youthful Ganymede to the sky to serve as his master's cup-bearer. The skywatchers of the Euphrates Valley first identified this area as an eagle.

Capricornus—The Sea Goat (cap-ro-*kor*-nis)

Culminating at midnight in early August, Capricornus, though filled with faint stars, is easy to find by joining the three brightest stars of Aquila with a line that extends southward. Long thought of as a goat, the Platonists thought of it as the "Gate of the Gods" for human souls to use on their way to Heaven. This gate was the opposite of Cancer, the "Gate of Men," which souls passed through on their way to Earth.

Cygnus—The Swan, the Northern Cross (*sig*-nuhs)

Looking much like the outline of a swan, or even a cross, Cygnus is one of the Northern Hemisphere's most famous constellations. It's a very busy place, for here the northern Milky Way is at its best. In fact, if you are under a dark sky you should see the Milky Way divide into two streams in Cygnus. A dark nebula between us and the more distant stars causes this apparent divergence.

Deneb is the Swan's brightest star. One of the mightiest stars known, it is some 25 times more massive and 60,000 times more luminous than the Sun. Some 1,500 light-years away, Deneb is by far the most distant star of the famous Summer Triangle; Vega is 25 light-years away, and Altair only 16.

The Swan is Cygnus's traditional name. One beautiful legend pictures Cygnus as Orpheus, who entered the sky as a swan so that he could be near his musical instrument, Lyra, the lyre or harp. In a very different legend, Zeus assumed the form of a swan in his seduction of Leda of Sparta.

Hercules—Zeus's Son (*her*-cue-leez)

Although Hercules has no prominent stars, its "keystone" of four stars is easy to find. The constellation lies within a triangle formed by Vega, Altair, and Alphecca, the brightest star in Corona Borealis. To the Phoenicians, this figure was the god Melkarth; the people of the Euphrates Valley saw the figure as a god of the Sun. The son of Jupiter and Alcmene, Hercules showed his strength and courage in his twelve labors. We have read of his second labor in the description of Hydra. Hercules killed Hydra, sailed on the *Argo* with Jason, and for good measure, helped to plunder Troy.

Lyra—The Lyre, the Harp, the Vulture (*lie*-rah)

Were it not for Vega, the constellation of the harp would be quite difficult to find. Lyra occupies a small portion of sky, and it consists mostly of a parallelogram of faint stars. But brilliant, blue-white Vega turns this group of stars into a real gem. The Greek legend about Lyra, the harp of Orpheus, is a sad story. Eurydice died on the day of her wedding to Orpheus, who begged Pluto to revive her. Pluto was so moved by his harp music that he agreed, so long as Orpheus never looked back. But just as Orpheus was leaving he did glance back, and so all he saw was his beloved Eurydice being swept back into Hades, never to return. The Chinese saw Lyra as a weaving princess, the Persians as a tortoise and later as a clay tablet. To the Romans, Lyra was a harp with bull's horns.

Ophiuchus—The Serpent-Bearer (oh-fee-*you*-cuss)

A large constellation covering a lot of sky, Ophiuchus is filled with some of the richest regions of the Milky Way. It culminates at midnight during the second week of June. A Greek word implying "a man holding the serpent," Ophiuchus was actually Aescula-pius, the "god" of medicine. He was so good at healing that he could even raise the dead, an action that offended Pluto, who persuaded Jupiter to put an end to his deeds and place him in the sky.

Constellations of Southern Hemisphere Summer (Figure 13)

Sagittarius—The Archer (saj-ih-*tair*-ee-us)

By far the most distinctive feature of Sagittarius is its teapot shape. Complete with pot, spout, handle, and steam that emerges as the faint band of the Milky Way, the asterism—the term used for any small group of stars—is a realistic one that is fun to look for. The handle also looks like a dipper, so some call it the Milk Dipper.

The ancient Arabs thought of the constellation as two groups of stars: the western part was a group of ostriches on their way to drink from the Milky Way, and the eastern quadrilateral, which is the Milk Dipper, was a group of ostriches returning from their refreshment. According to a Greek myth, Chiron, the civilized centaur, created Sagittarius to guide Jason and the Argonauts as they sailed on the *Argo*, which is probably why some charts show Sagittarius appearing as a centaur. The ancient Chinese saw the triangle as the House of the Winnowing Tray, used to clean chaff from harvested crops. The quadrilateral was the House of the Bushel, which was used to serve rations of food during times of hardship.

Scorpius—The Scorpion (*score*-pee-us)

A beautiful constellation filled with bright stars and a rich field of Milky Way, Scorpius also looks very much like a scorpion, com-plete with head and stinger. At the northern end is a line of three bright stars, with Antares at its center.

The "guardian of heaven" was the honor bestowed on An-tares, the brightest star in Scorpius, by the ancient Persians 5,000

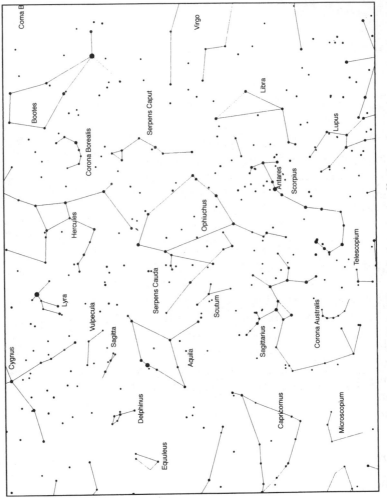

13. Southern summer constellations.

years ago. They thought of it as one of their few "royal stars." Since Sagittarius has been pointing his arrow at the scorpion almost since time began, one wonders how Scorpius has survived all these years. Guilty of poisoning Orion, Scorpius still seems able to wreak havoc on the countryside. It is possible that ancient Egyptians thought that since the Sun was in Scorpius during the coldest part of the year, perhaps they could blame the scorpion for causing their midwinter illnesses.

Scutum—The Shield (*scoo*-tum)

One of the sky's smaller constellations, Scutum is the home of one of the best star clouds of the Milky Way. Johannes Hevelius invented the constellation around 1690, and named it Scutum Sobiescianum or Scutum Sobieskii to honor King John Sobieski of Poland after he successfully fought back a Turkish army as it entered Kalenberg on its way to invade Vienna in 1683.

Constellations of Autumn (Figure 14)

Aquarius—The Water Bearer (ah-*kwair*-ee-us)

The Water Bearer is appropriately placed in the sky not far from a dolphin, a river, a sea serpent, and a fish. The ancient Babylonians saw the area as a man or a boy carrying a bucket. The relation to water is likely related to the Sun's crossing it during the rainy month of February.

Aries—The Ram with the Golden Fleece (*air*-ease)

The ancient Hebrews, Babylonians, Persians, Arabs, and Greeks were unanimous in seeing this constellation as a ram. In the Greek legend, the king of Thessaly's two sons, Phrixus and Helle, were abused by their stepmother. Mercury sent a ram with a golden fleece to help them escape. Although Phrixus made it to safety, Helle fell off the ram just as it was crossing the strait that divides

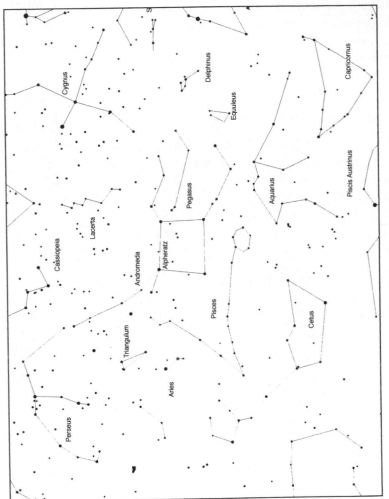

14. Fall constellations.

Europe from Asia, a body of water still called Hellespont. The sea of Helle is now familiarly known as the Dardanelles. Aries is traditionally known as the first constellation of the zodiac, since the Sun was at the "first point in Aries" on the day of the vernal equinox. However, the Earth's precession has made that moot, since the Sun is now in Pisces at the vernal equinox.

Pegasus—The Winged Horse (*peg*-ah-suhs)

Despite the fact that this constellation has no bright stars—nor are there bright stars near it—Pegasus is easy to find. Together with Alpheratz in the neighboring constellation of Andromeda, Pegasus's three brightest stars form the Great Square. Filling an otherwise empty region of sky, Alpha, Beta, and Gamma Pegasi, as well as Alpha Andromedae, are best seen in early September, when the constellation has its midnight culmination. According to Greek legend, Pegasus arose from the blood of the Gorgon Medusa after Perseus killed it. Perseus later used Pegasus to help rescue Andromeda.

Perseus—Rescuer of Andromeda (*purr*-see-us)

A striking constellation that culminates at midnight on November 10, Perseus appears as an arc of stars stretching from Capella in Auriga to Cassiopeia. The son of Zeus and Danae, Perseus went on an impossible mission to kill the monster Medusa. A single look at Medusa's awful countenance was enough to turn any creature to stone. Perseus slew Medusa, but kept her head in a bag as a souvenir. On his return, according to one version of the legend, he rescued Andromeda by showing the head to Cetus, who obliged by turning to rock.

Pisces—The Fish (*pice*-ease)

For thousands of years, many cultures have seen this constellation as fish, called either Pisces or Venus and Cupid. In the traditional Roman legend, Venus and Cupid were strolling along a Euphrates beach when Typhon, an evil dragon, appeared, ready to attack. The couple dashed for the water, turning themselves into fish. The stars in the western fish are called the Circlet. The ancient Chinese made a fence out of the Pisces region, calling it Wai-ping.

Constellations of Winter (Figure 15)

Auriga—The Charioteer (oh-*rye*-gah)

This pentagon-shaped figure is one of winter's most prominent constellations. It is led by Capella, the goat star, always accompanied by the three faint stars called the kids. Many ancient legends picture Auriga as a charioteer carrying a goat on his shoulder and kids in his hand. In another story, Erichthonius, the son of Hephaestus and Athena, invented a chariot to move his crippled body about. Capella has been the goat star since Roman times, rising with her three kids in November evenings to signal the start of the winter storm season.

Canis Major—The Greater Dog (*cay*-nis major)

Part of the sky's magnificent complex of winter constellations that includes Orion and his two dogs, Canis Major is best seen in early January, when it is in the sky almost all night. However, one of its responsibilities is connected with summer, when it also allegedly causes the hot and muggy "dog days" of September. In much of the Northern Hemisphere, Sirius rises at the same time as the Sun during the month of September, and legend has it that the Sun's energy is augmented by that of Sirius to produce the extra warmth. Mythologically, Canis Major is a serious hunting dog; with its many stars in a small space, it is also called a spotted hunting dog. The dog star, Sirius, is also the brightest star in the entire sky.

With its neighbor Canis Minor, the big dog is part of many beautiful legends. In one, the dogs squat under Gemini's table. The faint stars between Canis Minor and Gemini are supposed to be the crumbs that the twins are feeding the animals. But accord-

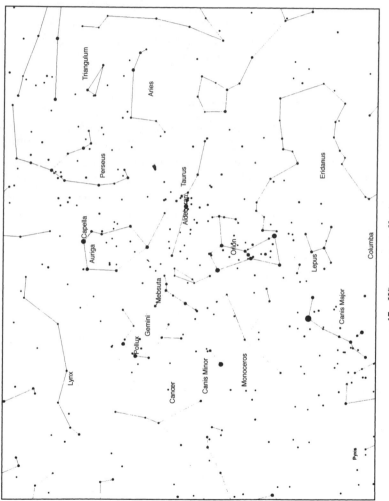

15. *Winter constellations.*

ing to the Greeks, Canis Major was not a dog to wait for crumbs. He was a racing dog who won a place in the sky after winning a race against a fox that was supposed to be the world's fastest creature. Jupiter put the dog into the sky to celebrate the victory. The most common legend, of course, has the dogs with Orion as he hunts. With his eye fixed on Lepus, the hare, crouching below Orion, Canis Major is ready for action.

Eridanus—The River (eh-*rid*-ah-nus)

This incredibly long constellation is somewhat hard to trace. Its source lies just west of Rigel in Orion, and then it flows southward. Its mouth is Archernar, a bright star so far south that most northerners never see it. The ancient Greeks gave the river its name, but for them, it extended only as far south as Acamar, or Theta Eridani, because observers in Greece could not see stars further south.

Gemini—The Twins (*jeh*-mih-knee)

With its two dramatic stars, Pollux and Castor, Gemini is easy to find, especially when it culminates at midnight in early January. It abides east of Taurus, northeast of Orion, and west of Leo. Different cultures have seen the twins as gods, men, animals, or plants. The Greeks called them Castor and Pollux, and used their names for the constellation's two brightest stars. They sailed with Jason and the Argonauts. Remembering how the twins helped save the *Argo* from sinking during a storm, sailors in later times considered the constellation a good portent. In 1781, William Herschel discovered Uranus near Eta Geminorum, and in 1930 Clyde Tombaugh discovered Pluto near Delta Geminorum.

Lepus—The Hare (*lee*-pahs)

Although Lepus has no bright stars, its position due south of Orion makes it a cinch to locate. The Arabs thought of it as Orion's chair. Egyptian observers dubbed it the Boat of Osiris. The Hare comes from the Greeks and Romans, who gave it the name Lepus. Since Orion liked hunting rabbits, it seemed appropriate to entice him with one. The constellation culminates at midnight in the middle of December.

Orion—The Hunter (oh-*rye*-on)

A real showpiece, Orion is without a doubt the most prominent feature in the Northern Hemisphere's winter sky. Orion has been recognized as a hunter for thousands of years. In Job 38:31, God asks if Job knows how to "loose the bands of Orion." To the Chaldeans it was Tammuz, after the month in which the familiar belt stars had their first rising of the season before sunrise, an event known as "heliacal rising." The Syrians called it Al Jabbar, the Giant. To the ancient Egyptians it was Sahu, the soul of Osiris, the god of light. The Bororo tribe in central Brazil saw the figure as Jabuti, the turtle. People in the Bahamas and other places picture it as Jacob's coffin.

In one Greek legend, Orion was the best hunter in the world, and he knew it. Angered at his pride, Gaia, the Earth goddess, sent a scorpion to bite him in the foot and kill him. Diana, the Moon goddess, then placed him in the sky as far as possible from Scorpius, so that he would never again be bitten. From March to May we see the drama replayed each night as Orion sinks to the Earth in the west while Scorpius rises in the east, and the following night, Orion, now restored to life by Ophiuchus, rises again.

Taurus—The Bull (*tore*-us)

A prominent northern sky constellation, Taurus culminates at midnight in early December. Taurus has been seen as a bull since the time of the Chaldeans some 5,000 years ago, and to the ancient Hebrews as well. To the Egyptians of 1700 B.C. it was Osiris, the god of the Nile. To the Romans, the bull was Zeus disguised. In one story, Zeus planned to kidnap the beautiful Europa, daughter of the king of Sidon. The malfeasance occurred as Europa was frolicking on a Mediterranean beach with friends. She noticed a handsome white bull grazing among her father's herd and climbed on his back for a ride. Europa got more than she bargained for. The white bull was actually Zeus in disguise. Leaping to its feet, the bull took off into the sea and swam to Crete.

The Pleiades

The seven daughters of Atlas and Pleione are the brightest stars of the most famous open star cluster in the entire sky. In a Native American tale, the Pleiades were a group of seven youngsters who, on a walk through the sky, lost their way and never made it home. They have stayed in the sky, close to each other so that they would not become separated. The seventh sister is hard to see because she really wants to go back to Earth, and her tears dim her luster. In China, they were the seven sisters of industry. Ancient Hebrews called them Sukkot R'not, for Tents of the Daughters.

On a reasonable night, you should be able to see at least six of the Pleiades, probably a seventh was a little brighter in ancient times. Under good conditions, you might see as many as nine. Containing more than 500 stars in all, the Pleiades is some 400 light-years away.

Southern Hemisphere Constellations (Figure 16)

Carina—The Keel of the Ship *Argo* (ka-*ree*-nah)

Attracting celestial sightseers and professional astronomers alike, Carina is one of the most beautiful regions of the Milky Way. It had a modest beginning as the keel of the ship that Argo built for Jason and his Argonauts as they sailed out to search for the golden fleece. The original Argo Navis was so vast, covering so much of the sky, that astronomers divided it into the Stern (Puppis), the Sail (Vela), and the Keel (Carina).

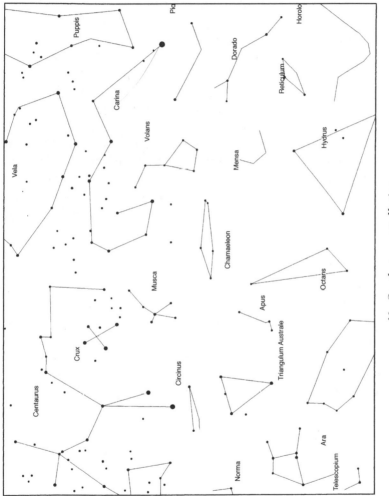

16. *Southern constellations.*

Puppis · Pla · Horolo · Dorado · Reticulum · Carina · Volans · Mensa · Hydrus · Vela · Chamaeleon · Musca · Apus · Octans · Crux · Triangulum Australe · Centaurus · Circinus · Norma · Ara · Telescopium

Centaurus—The Centaur (sen-*tor*-us)

This group of stars represents Chiron, a rare example of an upright centaur in an otherwise heinous and ruthless group of creatures. The centaurs were men down to their loins, and horses the rest of the way. Things started out well when the centaurs were all invited to the wedding of Pirithous and Hippodamia. Unfortunately one of the centaurs had too much to drink and threatened the bride, and then the other centaurs tried to do the same. A big fight broke out and, after that the centaurs never lost their reputation as a wild and mean bunch.

Chiron and Pholos were the two good centaurs. Chiron was their philosopher and educator, and after his pupil Hercules accidentally wounded him with a poisoned arrow, Chiron renounced his immortality and Jupiter launched him to the sky. In another version, Pholus, another centaur, is represented as the centaur in the sky.

Crux—The Cross (crucks)

The most famous southern constellation, the Cross was vital to sailors for centuries for a simple reason—it pointed the way to the south celestial pole. It is a part of the flags of several southern nations. Although it appears so obviously as a Latin cross, Crux was not mapped as a separate entity until 1592. In addition to a cross, the famous four stars have been pictured as such ordinary things as a knee protector, a net, a fishing spear, and an eagle's foot.

Hydrus—The Male Sea Serpent (*high*-drus)

The male version of the sea serpent is much more recent than the mythological female version. Johann Beyer first noted it in 1603, placing it near Archernar, the mouth of the river Eridanus. In one story, if Hydrus wishes to visit his mate, he swims all the way along the river Eridanus, from Archernar to the western border of Orion. Then he makes the risky land crossing between Orion and Lepus and finally crosses over the Milky Way. He has to do this quickly to avoid being missed in his usual place.

17. *Magpies across the Milky Way. Courtesy Project ACCESS.*

ACTIVITY: CONSTELLATION VIEWERS

OBJECTIVE: To get an idea of what constellations might look like in the sky.

MATERIALS:

Tubular cans meant for potato chips (e.g., Pringles)
Pattern sheets
Hammer(s) and nail(s)
Black butcher or construction paper circles the size of the can lid
Black butcher paper for wrapping the can

Gummed stars, crayons, or markers
Clear Contact paper (or patterned Contact paper)
Push pins
Scraps of Styrofoam or plywood
Tape or glue, scissors
Optional: white pencils, crayons, or chalk

PROCEDURE:

1. Using the hammer and nail, punch a hole in the center of the metal end of the can. This is the end you will view through.
2. Trace circles the size of the inside of the can lid onto black butcher or construction paper. The circles should be approximately 3 inches in diameter.
3. Choose a constellation pattern. Lay it on a black paper circle on the Styrofoam or plywood. Use a push pin to transfer the star pattern onto the black paper circle. The holes made by the pin should be as tiny as possible for maximum effect.
4. Optional: Use a white pencil, crayon, or chalk to connect the stars into the constellation "picture."
5. Cut a piece of black butcher paper to fit around the can (the Pringles can measurements are 23.3 cm by 23.7

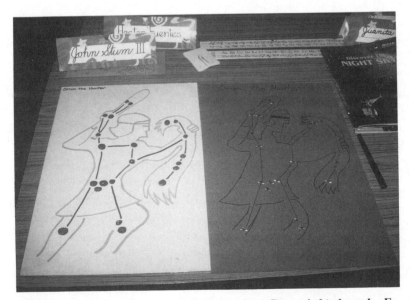

18. *Constellation patterns are studied in Joan Regens' third grade, Esperanza Elementary.*

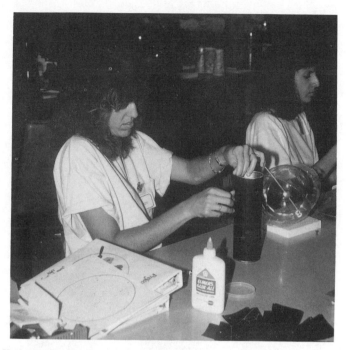

19. *Kari Richardson (left) and Velma Paez demonstrate constellation viewers.*

cm around, or 9⅜ by 9⅝ inches). Tape or glue the black paper to the can.

6. Use gummed stars or patterned Contact paper to decorate the can.

7. Optional: If you use gummed stars or drawings, cover the decorated can with a piece of clear Contact paper to protect the artwork.

8. Place one of the constellation circles into the plastic can lid, wrong side against the lid. Put the lid on the can. Hold the can up to a light and view the constellation through the nail hole in the metal end.

9. Store the constellation circles inside the can lid afterward.

20. *Rick Hoffman demonstrates a constellation viewer.*

10. Optional: Students can trade viewers and try to identify each other's constellations.
11. Optional: Use the smaller patterns with the cardboard tubes from toilet paper or paper towels. Punch out the patterns onto 4–5-inch squares of black paper. Fold the paper over the tube, right side of the constellation to the inside. Rubber band the paper around the tube.

ACTIVITY: SUN, MOON, AND STAR STORY BOXES

OBJECTIVE: To understand some Native American interpretations of the sky. Students will explore the different artistic views of the Sun, Moon, and stars as expressed by ancient and contemporary cultures. They will retell a myth or legend for themselves or to peers. This activity provides practice in storytelling, sequencing, listening, and comprehension.

STORY SUMMARY: Once there was a curious coyote who lived with his friends, the five wolf brothers. Every day the wolf brothers and their dog would go hunting. When they came home, they shared their meat with Coyote. They also talked around the campfire about something strange and frightening they had seen in the sky. But they would never tell Coyote what it was.

Every night Coyote would ask the wolf brothers what they had seen in the sky. Finally one of the wolf brothers said, "Let's tell Coyote what we have seen." They told Coyote about two strange animals they had seen high in the sky. They were very brave hunters, but there was no way they could get near the creatures. Soon Coyote had a plan.

Coyote gathered many arrows together and began shooting the arrows into the sky. The first arrow stuck. The second arrow stuck to the first, and the third arrow stuck to the second. After a while, there was a trail of arrows leading up to the sky.

The next morning Coyote, the five wolf brothers, and their dog climbed the narrow trail. They climbed for many days and nights, and finally reached the sky. The two animals in the sky were fierce grizzly bears, and Coyote was afraid. But the two youngest wolf brothers were not afraid. They approached the grizzly bears and nothing happened, so the next two wolf brothers followed. Finally the oldest wolf brother and his dog joined the group.

Coyote admired the beautiful picture they made in the sky. He began to back down the trail of arrows, breaking off the arrows as he went. To this day, the wolf brothers and their dog face the

21. *Lori Gaither shows how Coyote arranged the sky. SAMEC regional workshop.*

two grizzly bears in the sky. We call this sky picture the Big Dipper. When Meadowlark sings at night, he is telling everyone to come and look at Coyote's picture in the sky.

The grizzly bears are the stars in the Big Dipper's bowl that point to the North Star. The youngest wolf brothers are the stars that face the bears across the bowl of the Dipper. The middle wolf brothers are the first and last stars of the handle of the Dipper. The oldest wolf brother and his dog are Mizar and Alcor, the two stars that appear as the middle star of the handle.

MATERIALS:

Box, basket, or other appropriate storage container
Patterns (for tracing) or pictures of main characters

Scraps of cloth, felt, oaktag
Scissors
Glue

1 large piece of fabric (about 1 foot square)	Plastic or wood figures, if appropriate

PROCEDURE:

1. Present grade-appropriate factual information on the Sun, Moon, and stars, based on the appropriate chapters 7, 8, and 13. Share an astronomical myth, legend, or folktale such as "How Coyote Arranged the Night Sky."
2. Assemble a story box, as described below.
3. Discuss the origin of the story (historic time, location) and the differences between scientific information and legends.
4. Allow the children (singly or in small groups) access to the story box so that they can use the contents to retell the story.
5. Stress that the stories were important to their originators and that the story boxes should be treated gently. All small pieces should be wrapped in the background cloth and stored carefully in the box or basket.
6. Allow other options for responding to the story; e.g., drawing, painting, using clay, or writing new legends.

ASSEMBLING THE STORY BOX:

1. Choose a sturdy shoebox, basket, or other container for storing the background cloth and small pieces.
2. Represent key figures and objects in the story with felt cutouts, pictures glued to tagboard, or small figurines or recycled items.
3. Choose a piece of fabric to serve as a surface for the story figures. Wrap small pieces in this background cloth. Try to choose a color and type of fabric appropriate to the story. A piece of fabric about 1 foot square works well; however, the size really depends on the sizes and the number of pieces included.
4. Assemble the figures and objects, wrap them in the background cloth, and place them in a sturdy box. Label the box with the name of the story (or with representative pictures, if students are not yet reading).

ACTIVITY: THE BIG DIPPER STORY WHEEL

BACKGROUND: The circumpolar constellations are the basis for this story from eastern Canada. It is important to emphasize to the students that the people who told this story were describing the *apparent motion* of the Big Dipper and other star patterns around Polaris, the pole star. We now understand that it is the *motion of the Earth* that allows us to see the canopy of stars in different positions from hour to hour through a night and from month to month during the course of a year. Remember that there is a season for storytelling in some Native American traditions, which lasts from late fall to early spring. Traditionally, storytelling is not done at other times of the year.

THE STORY: Many years ago people looked into the night sky and imagined wonderful stories in the stars. One story lasts for a whole year, and tells about the adventures of the Great Bear and the Bird Hunters.

When winter ended, the Great Bear left her cave. She was very hungry after her long sleep and was anxious to find food. But hunters were following her.

Seven brave Bird Hunters followed the Great Bear across the sky. Robin led the hunt, followed closely by Chickadee and his cooking pot and Moose-bird. Farther behind were their friends: Pigeon, Blue Jay, Horned Owl, and Saw-whet. The bear looked big and clumsy, but she moved across the sky quickly. The hunters followed behind all summer, but as autumn approached they still had not caught up to the Great Bear.

Some of the hunters became tired and discouraged. Saw-whet, the last hunter in line, left the hunt. Soon Horned Owl also gave up and went in search of Saw-whet. Blue Jay and Pigeon tried to keep up with the leaders, but soon they also left the hunt and flew home.

Only Robin, Chickadee, and Moose-bird followed the Great Bear into the autumn. The bear grew angry and rose up on her hind legs. She growled loudly and clawed the air to scare the hunters. But Robin was a brave hunter. He shot an arrow and hit

22. *Anita Mendoza and the Never-Ending Bear Hunt. Project ACCESS!*

the Great Bear. Drops of her blood fell on Robin's shoulders, turning his breast a bright red. Other drops fell on the autumn leaves, coloring them a bright red.

When winter came, the dead bear lay on its back up in the sky. But her spirit returned to the cave and entered another bear. In the spring, the new bear will leave the cave again to travel across the spring and summer sky, always pursued by the Bird Hunters.

OBJECTIVES: The students will learn about a Canadian Indian tale and will compare the action in the story to the scientific facts about the motion of the Earth and apparent motion of the stars.

MATERIALS:
"Never-Ending Bear Hunt"   Ruler
    story, patterns, template   Stapler

1 large sheet construction paper per student
1 paper plate per student
Glue or rubber cement

Markers, crayons, paper scraps for decorating
Brass fasteners

PROCEDURE:

1. Discuss with the children the ideas of legend and science.
2. Share the story of the "Never-Ending Bear Hunt" with the students. You may want to bring in details from your

23. *It is easy to spot bright constellations. Cygnus, the swan (or Northern Cross), sprawls from upper left to center. Lyra, the harp, is at right. Photo by David Levy.*

discussion from step 1: Discuss seasons, patterns of stars, the Big Dipper, Polaris, and even navigation by the stars.

3. You will need one paper plate (sturdy ones with raised edges do not work well), one pattern sheet, one brass fastener, and one large piece of construction paper for each student. Crayons or markers for decorating the constellations and/or the construction paper foreground, glue, and a stapler are also needed.

4. Decorate the constellation pictures on the circular pattern. Cut it out and glue it to the center of a paper plate.

5. Using the large template (or the measurements from the template), locate and mark the position of Polaris on a large sheet of construction paper. Option: Prefold the paper and mark the positions for Polaris and the side staples for younger students.

6. Push the brass fastener through Polaris on the paper plate pattern and again through the mark on the construction paper. Secure the fastener.

24. *Magpies across the Milky Way. ARTIST 1994.*

25. *Larry Lebofsky and Betty Fulcomer show constellation viewers to a second-grade class.*

26. *Larry Lebofsky and Betty Fulcomer's second grade. Robison Elementary.*

7. Fold the construction paper up from the bottom edge as indicated. Staple along the side edges, close to the edge. Be sure the staples are not too close to the top (horizontal) folded edge; the paper plate needs clearance in order to rotate, as Polaris is not exactly in the center.

8. Slowly rotate the paper plate counterclockwise, allowing the bear to exit her den and the hunters to fall below the horizon, as described in the story. If the horizon is too low, the students can add trees, rocks, or other decorations to raise it. Continue to decorate the construction paper foreground with crayons, markers, construction paper scraps, etc.

9. Share the story again. The teacher or students can rotate the paper plate to illustrate the changing positions of the constellations throughout the seasons.

10. Emphasize again that it is not the stars that move, but the motion of the Earth that makes them appear to move. Early cultures used star positions and stories to mark seasons, especially planting and harvesting times, as well as stories to teach lessons. We can value these stories as literature, even though the science in them might not have the accuracy of our current knowledge.

Part II

Getting Ready to Observe the Sky

Chapter 4

Keeping an Observing Log

"What are you doing now? he asked. 'Do you keep a journal?' So I make my first entry today." Henry David Thoreau penned these words in 1837. One of America's great writers, and not coincidentally one of the most astute observers of nature who ever lived, Thoreau made a point of observing the physical world around him and keeping a journal in which he recorded what he saw. Thoreau thought deeply about his experiences. "Both place and time were changed," he wrote of one of his first nights under the stars, "and I dwelt nearer to those parts of the universe and to those eras of history which had most attracted me."

More than 200 years ago, Charles Messier, a French astronomer and the first person to conduct a successful search for comets, kept a detailed journal of his nights under the stars, a log in which he noted what he observed, the weather conditions, and his complaints about how his telescope needed repair and upgrading. His log contained all the thoughts and concerns that are shared by people who look at the sky today. When one of us (David Levy) visited the Paris Observatory in 1986, he found a page of this observing log displayed under glass. In a very special way Levy felt, like Thoreau, that both place and time were changed. He felt as though he were right there beside the great scientist. Thanks to that observing log, the two observers were together, about to sit down and chat about their love of telescopes, comets, and sharing the sky with others.

Such is the wonder of keeping a record of your nights under the stars. But there are practical reasons for log-keeping. Seeing the sky is a process that involves the eye and the brain. Recording what you see enhances this process because it helps your brain interpret what you see, resulting in better observations and a better memory of these observations. Writing things down brings order to your observing experience, and it can increase the fun you have while observing.

Recording is not just something to do; it is a way of becoming a more observant person. Encouraging your children to record what they see will help them become better watchers of the world around them. What they write about their observing sessions is much less important than the fact that they do record them. In this chapter we suggest a format for recording, but this is only a guide; teachers, parents, and children alike should feel free to develop their own ideas.

Observing logs can be as simple as a short listing of the date, time, and objects viewed, or it can be as complex as one containing artwork and narrative. It is probably best to start with something basic. David Levy began recording his observing sessions at age 13, during the summer of 1961. He began his record thanks partly to the encouragement of his seventh-grade science teacher, Mrs. Forsythe, who understood that the sky was destined to play a big role in her young student's life. Fortunately he had a good memory of the place, date, and time of Session 1 that had taken place 2 years earlier, as it had involved a partial eclipse of the Sun on the morning of October 2, 1959. He remembered the few summer sessions he enjoyed with his first telescope, looking at Jupiter during August 1960, and within a few hours he had brought his observing log up to date. By the time he was in high school, Levy's observing log had grown so large that it stood on its own as a successful science fair project. In early 1997 he recorded his 10,000th observing session, including four total eclipses of the Sun and 21 discoveries of new comets.

KINDS OF RECORDS

Notebook

By far the most common type of record is a notebook in which you record your time under the sky. For children, the simplest kind of notebook is a *Moon journal*, a booklet designed to record the nightly changes of the Moon's phase and its position in the sky. The idea of such a journal is to introduce the child to basic observing and recording techniques by keeping the recording simple and directed toward one object, such as the Moon. In the ARTIST program, teachers and students recorded their observing experiences in a variety of ways. Some drew a rough circle or crescent to denote the position of the Moon each night, with no record of the time, the phase, or an accurate rendition of the Moon's position in the sky. Others were more ambitious, recording times, places, and phases accurately. Logs included poetry, artwork, and anecdotes. We think that children should be encouraged as early as possible to keep a full observing log that incorporates observations of the Moon's phase, its position both in the sky and relative to the skyline, notes on other objects, and weather conditions.

The notebook can have almost any type of binding, so long as it is of good quality. In humid weather spiral bindings might not keep the pages intact, and just turning the pages could cause them to fall out. Stapled and glued bindings probably work better.

The example that follows is based on David Levy's observing log. Though it has worked for him, it is only an example. Observing logs can be as simple, or as complicated, as teachers, parents, and students wish.

Data

When Levy began his observing log at age 12, he included the following material:

- Date
- Time started
- Time ended
- Weather conditions
- Place
- Telescope
- Others present
- Objects viewed

Even though Levy developed this coded system when he was a child, he has used it with little change for 37 years. It is far more inclusive than many observing logs, especially children's logs, need to be. Although you can design your own log based on specific needs, this one does serve as one successful model. Here is an example, taken from an important session that took place during the morning of April 15, 1994. Notice that an observing log can be personal and fun:

(SESSION NUMBER/DATE/TIME/WEATHER/PLACE/TELESCOPE/
OTHERS PRESENT/OBSERVATIONS)

*9202/April 14–15/0345-0600/10-f/Jarnac/Miranda/
—/CN3 30 minutes, eastern sky.*

Discovered a comet near Alpha and Beta Equulei. Found just before 4 AM after about 10 minutes of comet hunting, and while talking over the observatory phone with my friend Peter Jedicke. Since my computer was in the shop, I had to resurrect an old computer to find out that the comet was a new one—what a feeling to find out it was! Afterward I walked up and down the street. Is it possible that the birds really are chirping more excitedly this morning?

*Comet magnitude 10.5. Position in sky: Right ascension 21 23.4
Declination +5 26.*

SESSION NUMBER: This number is consecutive. This was Levy's 9,202nd session since the solar eclipse of October 2, 1959. It is interesting to start keeping records early in life, and then keep on doing it!

DATE: Because the date changes at midnight, Levy records the night of April 14 and 15 as a double date.

TIME: The time the session began, 3:45 AM, and the time it ended more than 2 hours later are recorded here.

WEATHER CONDITIONS: Levy's original scheme included an 11-point weather description, where each number represents a distinct sky condition that significantly affects the observations made. The number 10 means that the viewing conditions were perfect. The letter f (for favorable) means that Levy was satisfied with the session. Other comments, about high wind or biting cold, for example, could go here. Here is the observing conditions code that Levy uses:

Cloudy sky

0: Rain, snow, or other precipitation. While very few sessions have a 0 in their weather list, the ones that do are interesting. In 1985 Levy waited through driving rain at an observatory where he hoped to obtain an image of Halley's Comet. The rain stopped and a hole in the clouds revealed the comet for less than 5 minutes. The resulting picture showed the first eruption of a jet of dust from the comet in three-quarters of a century.

1: Sky completely cloudy.

2: Sky mainly cloudy, with occasional clear patches through which some observing might be possible from time to time, if you're lucky.

3: Partly cloudy sky. Many nights are like this, either with cumulus clouds passing over or patchy layers of stratus or cirrus clouds.

Hazy sky

4: Sky so hazed over that observing is difficult. Particularly humid summer nights are often so hazy that only the brightest planets and stars are visible.

5: Hazy sky. Humidity or dust particles are causing reduced visibility.

6: Slightly hazy sky.

Clear sky

7: A typical clear sky in a city, with some light pollution reducing visibility.

8: Suburban clear sky, with fainter stars and the Milky Way visible.

9: Very clear sky, a condition not seen unless you are miles from a city. The Milky Way is clearly visible when it is significantly above the horizon.

10: Absolutely fantastic night, as session 9202 was. A sky so clear that its velvety blackness is pierced only by myriad stars and the Milky Way.

PLACE: The observing site is recorded here. Jarnac means that Levy was at his Jarnac Observatory.

TELESCOPE: The size and type of telescopes or binoculars are recorded here. Miranda is the name of a 16-inch-diameter reflecting telescope.

OTHERS PRESENT: A list of others who observed with you belongs here. Sometimes your guests write their names and messages. Levy was observing alone on the morning of April 15.

OBSERVATIONS: Here is a list of what was observed. You could include drawings of the Moon or planets here, plus comments about what the observed objects looked like and how they compare with earlier views. CN3 is yet another code, this time the designation for Levy's comet hunting project, and Levy recorded that he discovered a new comet that morning. The numerical information at the end includes the comet's brightness; magnitude 10.5 means that it was easily visible through a moderate-sized telescope, 6-inch diameter or more, but much too faint to be spotted with an unaided eye or even powerful binoculars. The right ascension and declination refer to the comet's position in the constellation of Equuleus, the little horse. Equuleus is one of the smallest of the constellations.

Narratives

Young people often prefer to write up their experiences in the form of a story or poem. Since this method helps develop writing skills in addition to proficiency in observing, it should be encouraged, especially for elementary school children.

Art

Young children enjoy expressing what they see and what they think in the form of artwork. Using the observing log as a portfolio of their voyages into the night sky is an excellent idea, although the children should note the difference between what they see in the sky and what they imagine there.

Photography

If you or the children take pictures either of the sky or of the others at your observing session, these should be included in the observing log. Together with the observations, a narrative, and notes from friends, this summary of the night's experience should provide a memory of a productive and enjoyable evening.

Tapes

Recording by audio or video tape is a modern way of keeping notes on what you have observed. We have recorded several observation sessions this way and have gained some precious memories. What better way to remember the December 1963 eclipse of the Moon? On that night a group of young Canadian high school students recorded the times during which the Earth's shadow crossed over lunar craters and marked the end of the event with an enthusiastic rendition of Canada's national anthem.

Audio taping is a preferred means of recording a group that observes a shower of meteors (see chapters 10 and 11). Children can simply announce their sightings, giving all the necessary information, and the tape recorder will record it for later evaluation.

Videotaping an observing session will yield a record of the night for the purpose of evaluating the children's participation or showing others how much children can enjoy an evening under the stars. But use videotaping sparingly: at night, the same light you need to record the children looking at the sky will seriously hinder their ability to see anything under that sky. It is also possible to attach the video camera to the telescope to record the image of the Moon or planet.

Young computer enthusiasts can record their sessions on computer, either in a simple text file or using a database program. With such a program, they could quickly look up when, for example, they last saw Saturn.

PRIDE, PERSPECTIVE, AND POSTERITY

If the idea is introduced with enthusiasm and care, keeping an observing log could become a pleasurable activity. The attention that children pay to detail in keeping a log over a long period of time should instill a sense of pride and an understanding that observing is a special activity that has lasting value. This activity also encourages observers of any age to keep their enjoyment of the sky in perspective. If observing sessions follow a routine that gets written down, then children will remember more from that routine.

Posterity—both from personal and general points of view— is a good reason to keep a log. Just as Charles Messier's carefully kept observing logs have survived for two centuries, the ones that you and your young charges record tonight could have some redeeming value in the future—for themselves, for their children, or for some other unforeseen purpose. What happens if you see, for example, a very bright meteor, one that turns out to have survived its plunge through the atmosphere and hit the ground? If you have carefully recorded your observation, it might be very valuable to pinpoint the path of the object and help find where it landed. Many unrecorded discoveries have soon been forgotten, but history is full of recorded notes of seemingly trivial events that subsequently turned out to be crucial.

Keeping a log is a way to encourage the children to take the sky seriously. Anything worth doing, they should know, is worth doing well, and a night spent looking at distant places and far-off times is a night that will not be soon forgotten. Observing the sky at night is an experience that transcends the usual. It lifts us from the ordinary events of the day to a higher plane, and recording the experience in an observing log helps us feel a part of the night sky and the wonders it offers. "Follow your genius closely enough," Thoreau advised children of all ages, "and it will not fail to show you a fresh prospect every hour."

Chapter 5

Telescopes for Children

"Which telescope shall I buy for my child?" We have heard this same question many, many times, and it is a good one. It is also a very difficult one to answer.

David Levy's first telescope was a small 3.5-inch f/10 Skyscope, bought at New York's Hayden Planetarium gift shop in the mid-1950s. While visiting there, his uncle saw it and thought it would be just the ticket to get the children interested in astronomy. By 1960, the telescope was still little used, but one memorable afternoon Levy walked into his living room to see his uncle and his father figuring out the telescope. The night that followed was golden. Two bright "stars" were in the southern sky: The family set up on the brighter one. They quickly learned about focusing. By changing the distance between the telescope's main or primary mirror and the eyepiece (they accomplished this by moving the eyepiece in and out), what began as a bright light quickly shrank to a beautiful disk with two belts across its center and four small stars nearby. That bright star, it turned out, was the giant planet Jupiter.

You cannot do much better than getting your first look at the sky through your very own telescope—*if the telescope is a good one.* It is also possible that a good pair of binoculars will work just as well.

WHY NOT START WITH A PAIR OF BINOCULARS?

With their ability to magnify the heavens to a small degree, binoculars can show the sky favorably. A pair of binoculars is really a combination of two low-powered telescopes joined together so that you can use both eyes instead of just one. Binoculars will show you craters on the Moon, the moons of Jupiter, and more than five times as many stars as you can see with the naked eye.

The single biggest problem with binoculars is holding them steady. If you sit on a lawn chair, you can use the armrests to provide support for your elbows as you hold the binoculars. This way they will provide views that show the stars as sharp and steady points of light rather than as fireflies dancing about the field of view.

The way binoculars work depends on two things, the diameter of their objective lenses as well as the magnification of their eyepieces. The different combinations result in many possibilities. For ideal night viewing, we recommend using 7 × 50 binoculars. The 7 refers to the magnification, and the 50 is the diameter of each of the objective lenses in millimeters (mm). The preferred shipboard size in the past, these "Navy night glasses" use wide objectives to gather large amounts of light. However, they can be expensive, $200 or more. A well-made pair of binoculars with 20- or 30-mm lenses would do well for children, for example, an 8 × 30 pair.

Unfortunately, many cheaper binoculars found in department stores are useless! The reason is that unless the two small telescopes that make up the binoculars are precisely aligned with each other, each eye will see a slightly different view. At first you won't notice this because your eyes will try to correct the problem. However, you will quickly find that viewing becomes uncomfortable as your eyes strain to merge the two images, and you might get a headache.

You can do one or two quick tests at the store to determine whether or not the two lenses are collimated—in other words that they are looking at precisely the same field. The simplest test is to have someone hold a book or some other object over one of the objectives while you look at the distant object. Both your eyes should be open. Then ask that the book be moved away quickly. If you see two images at first, which then merge, the binoculars are out of alignment. In another test, look at the top of a distant wall or at the roof of a building. Then as you continue to look, move the binoculars until they are about a foot from your eyes. As you move the binoculars outward, the single image will break into two images. If the binoculars are out of alignment, one roof or wall top will appear not to be in line with the other.

*N.B. Do not use your binoculars to observe the Sun, unless you have proper protection for your eyes, as **permanent blindness** can result from even a quick look through binoculars.*

Children should enjoy looking through a pair of binoculars, and with reasonable care (especially warning the children not to drop them) a pair should last indefinitely. The Moon is an ideal target for observation: Binoculars will reveal the Moon's mountained and cratered surface. However, there soon comes a time when a child wants a more powerful view, and it is time to try a telescope.

CHOOSING A TELESCOPE

A common suggestion is that it's better to have a good pair of binoculars than a cheap telescope. We resist that idea. A telescope, once set on an object, can show the same view to a whole line of children. Besides, children enjoy the detailed views that telescopes offer. Instead of avoiding an inexpensive telescope, choose one that has been designed for beginners and that offers a good mount and good optics. We have found some fair telescopes—and some poor ones—that cost about $200. It is important to look closely at the telescope before you buy it. Does it have a sturdy base or does it wobble? Does it offer at least one good eyepiece? Telescopes use different eyepieces to achieve different magnifications. Although we admit that it is hard for beginners to tell in a store if an eyepiece is a good one, there is a simple test. The barrels—the parts that fit into the telescope—of the worst eyepieces usually have a diameter of less than 1 inch. A 1¼-inch-diameter eyepiece is probably a reasonable one. Although there are exceptions, this is a good rule to follow, and we strongly recommend that you avoid buying a telescope with an eyepiece with a diameter of less than an inch.

TYPES OF TELESCOPES

Levy's first telescope was a *reflector*, which uses a mirror to gather light and form an image. Starlight hits the mirror and goes back up the tube. Just before it converges, the light hits a small secondary mirror that directs the light out the side of the tube to the eyepiece. The first reflector was used by Isaac Newton in the mid-seventeenth century. The other common type of telescope is a *refractor*, a telescope with a lens at one end and eyepiece at the other. This type of instrument dates back to 1610, when Galileo first used one to look at the Moon and at Jupiter. Light from a star enters the lens and converges to a focus at the eyepiece.

A modern version of reflecting telescope uses a simple mounting made of plywood. Created by California amateur astronomer John Dobson, these Dobsonian mountings are sturdy, efficient, and inexpensive. However, the large telescopes (typical size is an 8-inch-diameter mirror) might be hard to move around.

Some stores sell more complex reflectors called Schmidt-Cassegrains, which are relatively compact because they re-reflect the incoming light. The main mirror has a hole in the center. Light from the sky hits the mirror, then hits a secondary mirror that sends it back down the tube and through the hole to the eyepiece. These telescopes are quite expensive, and we therefore recommend the two simpler kinds for children.

PURCHASING A TELESCOPE: SPECIFIC BRANDS AND RECOMMENDATIONS

Two generations ago, acquiring a first telescope was a simple task: you could grind your own mirror and construct the tube and mounting out of wood and pipe fittings. Or you could buy a small refractor with a 2- or 3-inch-diameter objective lens. When this book's authors started in astronomy in the 1950s and 1960s, the situation had changed. New companies had sprung up, offering to build reflectors for those who did not want to work the glass themselves. It was now possible to buy, complete and ready to use, a 6-inch-diameter reflector with a far greater light grasp than a 2-inch refractor, but *at the same price!* No wonder that skywatching was surging in popularity.

In the late 1990s, there is so much on the market that, as a beginner, you might be confused. What follows are our recommendations. There are many good telescopes out there, and, we hasten to add, we might have overlooked some manufacturers with very good products. We chose the following list based on what is most easily available. The quoted prices are for 1997.

THE MEADE MODEL 231 REFRACTOR. A 2.4-inch refractor that is easily available at Nature Company stores around the United States and at telescope and camera stores, this little telescope is fine for observing the Moon and bright planets. As with all telescopes, when you set it up for the first time, align the finder with the main telescope. Do this by finding a distant streetlight, bright star, or planet through the main telescope and then adjust the finder's screws so that the object is also centered in the finder. The telescope comes with a simple altazimuth mount (around $200).

STARGAZER STEVE'S 3-INCH SGR-2 NEWTONIAN REFLECTOR. This excellent 3-inch-diameter f/10 telescope was intelligently designed by Stargazer Steve Dodson (1752 Rutherglen Crescent, Sudbury, Ontario, P3A 2K3, Canada) specifically as a first telescope. It is a small instrument of very good quality. It has a wooden tripod and comes with a single medium-power eyepiece (around $200).

Steve notes the 3-inch is permanently collimated and there are no setup chores, no nuts and bolts or mechanical adjustments (aim and focus!). The step-by-step manual and video included with the kit ensure building and observing success as well as a deeper understanding of the scope and how it works. Stargazer Steve's quality hand-crafted 3-inch reflector is an excellent beginner's telescope—set on a smooth steady hardwood mount, it provides easy access to excellent sky views at a very reasonable price.

DELUXE REFLECTOR KIT BY STARGAZER STEVE. This is a telescope kit that comes with "everything you need to build your own mounted 4.25-inch telescope in a day, and observe with it the same night." Designed to be as easy to set up as some so-called prepackaged telescopes, this kit is strongly recommended.

EDMUND 3-INCH F/6 REFLECTOR. From Edmund Scientific (101 E Gloucester Pike, Barrington, NJ 08007) comes a compact 3-inch-diameter telescope. The f/6 means that the tube is more compact; the telescope brings light to a focus in a shorter space than the f/10 (about $230).

MEADE 390 REFRACTOR. A 3-inch refractor with altazimuth mount. This is a substantial telescope for a first-time user. Excellent for viewing detail on the Moon and bright planets, double stars, star clusters, and nearby galaxies (around $500).

MEADE 4500 REFLECTOR. In this price range, it is hard to decide which is better, the 390 refractor or the 4500 reflector. Both gather about the same amount of light and offer images of comparable quality. With the reflector you might need to adjust, or collimate, the mirror from time to time. The refractor lens is permanently mounted and should not need adjusting (around $400).

CELESTRON FIRSTSCOPE 114 DELUXE. Designed as a first telescope, this instrument has a mirror around 4 inches in diameter. In its deluxe version, which we recommend over the standard, the telescope comes with two 1¼-diameter eyepieces.

CELESTRON 4.5-INCH TELESCOPE WITH POLARIS MOUNT. Small telescope noted for a good-quality equatorial mount (about $600).

EDMUND 4-INCH ASTROSCAN 2001. Unique reflector telescope with ball-in-socket mounting. Felt pads on the socket support the wide red ball containing the telescope's mirror. Good for the Moon; excellent for low-power views of wide areas of the sky (about $350).

CELESTRON C-5 PLUS. A compact telescope with a motor drive, this instrument, though relatively expensive, is a very convenient telescope to use. It fits onto a tabletop. Once you set the polar axis on the pole, you simply turn on the drive, find the object, and the telescope follows the object *by itself*.

MEADE ETX. A recent addition to the telescope family, this instrument offers all the convenience of small telescope optics with the advantage of extreme portability (around $600).

DOBSON TELESCOPES: ORION 6-INCH DEEP SPACE EXPLORER (about $350), CELESTRON 6-INCH STAR HOPPER f/8 (about $450), and MEADE 6-INCH STARFINDER (about $450).

For those who want larger aperture, several companies now produce "big brute" 6- or 8-inch-diameter reflectors that you can buy at about the same price as the 3-inch and 4-inch traditional telescopes! Sturdily built, these telescopes are available because of the vastly simplified new technology for telescopes. Instead of a formal mount, they offer a plywood base that swivels on Formica and Teflon. There are no slow-motion controls, no motor drives—everything is simple. These telescopes are ideal for many observers, but they do require a little effort to set up. From a homemaker's point of view, they are not the most physically attractive telescopes.

The major astronomy magazines, like *Sky and Telescope* and *Astronomy*, contain the latest telescope information from many manufacturers and dealers.

AN OLD REFLECTOR AND A NEW IDEA

Most families who own a telescope have a cheap model they might have purchased at a department store. The trouble is that most of today's small department store telescopes are so poor that they do not give good views. When they visit schools, many astronomers bring fairly expensive telescopes with them. This way, children and their teachers get one good look before going back to their small, cheap telescopes with inadequate views. Result: either they realize their telescopes are worthless or they think that they do not know how to use a telescope. Either way, the small telescope, and the interest, are put in storage forever.

There has to be a better way to put even the cheapest telescope to effective use. But in most cases, the first contact a beginner has with a telescope is through a very poor eyepiece, with a tiny lens to look through. With that in mind, Levy brought out his trusted friend Echo, his first telescope, and did some work on it. He replaced Echo's 0.975-inch eyepiece holder with a larger one. The new focuser was 1¼ inches in diameter, the size chosen for most American eyepieces because in the heyday of amateur telescope-making, that was standard pipe-fitting size and people could make their own eyepieces easily. Levy inserted an eyepiece, looked, and was astounded to see the best view Echo had provided in 35 years. We now use Echo at virtually all our evening observing sessions at elementary schools, and thousands of children have already looked through its four-decade-old optics.

For viewing a bright object, it doesn't matter how good the telescope's mirror or lens is, so long as it is of long focal ratio, meaning, essentially, that the telescope has a fairly long tube. Most small beginner's telescopes are ground to f/10 or larger because

27. *Clyde Tombaugh, the discoverer of planet Pluto, poses with Echo, David Levy's first telescope.*

such mirrors can be produced relatively cheaply. If a 3-inch-diameter mirror is f/10, that means it focuses light at an eyepiece 30 inches away. A 3-inch f/5 sends light to an eyepiece 15 inches away. (Longer focal-length mirrors can be ground so that their curves are spherical; mirrors that are more highly curved require an additional process so that they are curved into parabolas.) Thus, an f/10 mirror, which is inexpensive to make, will provide images as good as a more expensive f/5 mirror, whose parabolized surface is more expensive to grind but allows the telescope tube to be shorter and lets the viewer see more sky at one time.

The faster a mirror is, the more sky it will show at one time.

But mirrors of shorter focus are more expensive. For a beginner's telescope, all that is required of the mirror or the objective lens is to grab some light and bring it into focus in a good-quality eyepiece.

With this approach, it is now possible to take the millions of cheap telescopes lying in closets around the world and put them to good use. Our suggestions for users of these readily available telescopes:

THE EYEPIECE HOLDER: If you have a refractor telescope, invest in a 1¼-inch adapter that will attach to the bottom end of the tube. This adaptor looks like the right-angle viewing device that comes with the telescope but has a fatter side that accepts larger 1¼-inch-diameter eyepieces. If you buy a small reflector, make sure it comes with an eyepiece holder that accepts 1¼-inch eyepieces. It is much harder to replace the small focuser with a larger one, as Levy was forced to do. Since the light is brought back out near the top of the telescope, there is no easily replaceable adaptor for a reflector. Instead, one must enlarge the hole at the top of the telescope tube, which will take a little drilling, cutting, and aligning. Finally, if you make this change, you must use what is called a low-profile focuser; otherwise the eyepieces might not get close enough to the tube to focus.

THE EYEPIECE: Now invest in a good-quality eyepiece. Besides their generally better optical quality, the bigger eyepieces have a major advantage because they have more glass. In school after school, we have found that the wider the eye lens, the easier time the children have finding where the glass is in the dark. Do not buy too high a power. Eyepieces rarely come with their magnifying power printed on them, since the power they give depends on the telescope you use with the eyepiece. Instead, the eyepiece has a focal length, such as 25 mm, printed on its barrel: The longer the focal length, the lower the power. We recommend an eyepiece between 30- and 20-mm focal length.

THE MOUNTING: Although one complaint against first telescopes is their wobbly mountings, a lot can be done to improve these mountings without spending an additional penny. Make sure that all the bolts are snug and then spread the legs out until the chain holds them tightly. This simple procedure will make the telescope much easier to use.

When we go to schools, we do use large telescopes. But we bring Echo along every time. This way, the children get an idea of what their own cheap, first telescope is capable of doing if used with a little tender loving care. They seem to enjoy knowing that the telescope Levy started with when about their age is still capable of breaking the bonds of Earth and flying to new worlds.

Chapter 6

School, Home, Camp: Differences in Approach

This book is written for teachers in various situations and for parents. In an important sense, anyone who has passed an hour under the night sky with a child is a teacher and should benefit from what this book offers. This chapter considers these special hours under the sky, whether they are held outside a classroom, at a summer camp, or in your own backyard.

THE SCHOOLYARD OBSERVING SESSION

Much as we desire it, an evening under the stars is a rarity for a school. Scheduling is a problem, and as often as not the weather will interfere. Also school grounds are not usually designed for observing sessions. They are filled with unshielded security lights, and often unexpected things happen, such as sprinkler systems going off unexpectedly during evening hours. However, with a little planning, it is possible to have an enjoyable evening session in a schoolyard.

The first rule is to *be flexible*. If you schedule the evening, be prepared with a system for informing the children and their parents if the skywatching session will be held or not. In most cases you will have a pretty good idea by midafternoon if the sky is going to be clear. However, there is a chance that a clear sky will cloud over at the last minute or that a sky cloudy in the afternoon will clear by evening. If the class is small, consider asking everyone to phone a number with a recorded message or phoning everyone yourself at dinnertime to tell them whether or not the event is on. If you end the school day with some uncertainty as to whether the skywatching session will take place and it does not, then someone should always appear at the site just in case a family did not hear that the event was canceled.

Preparing the Site

Before you gather a group of children and parents in a schoolyard, make sure that the site is ready for them. Will bright lights swamp the yard, preventing any real observing? Will the site be damp, muddy, or dusty?

On one clear night we had set up a telescope with a group of 25 children and parents. Suddenly a loud hissing sound interrupted the introductory remarks as several water sprinklers turned on. Startled and wet, the group scattered quickly. Although we did regroup nearby a few minutes later, the whole mishap could have been avoided had someone asked the school's maintenance department about the watering schedule. They might have be able to override the sprinkler timer and to turn off any annoying lights. The school might want to alert the local police department to avoid any unnecessary questions or interruptions. It is important to look into these situations before the skywatching session begins.

Alternative Activities

Thea Cañizo, a middle school teacher in Tucson, plans alternative indoor activities during skywatching sessions. When the group is large, some students can be working on other activities; this avoids long lines at the telescope and increases the richness of the evening activity.

Safety

More important than anything else is the idea that the observing session should be a safe and enjoyable experience for everyone. Know who is there, preferably by taking attendance in the classroom before going outside. Although we have rarely had any problems with party crashers, keep an eye out for gangs and other unfriendlies.

Skywatching Etiquette

A group of people under the night sky should know the following basic rules of conduct:

1. Allow the group at least 10 minutes to adapt their eyes to the dark.
2. While the stargazing session is under way, flashlights should be off. Children should know that shining a light in a friend's eye will interfere with the eye's adaptation to the dark.
3. Do not touch the telescope, especially when someone else is looking through it. The smallest tap will make the image jiggle, and the telescope could swing away from the object you're looking at.
4. The skywatching session is not "free play," and children should be discouraged from running around.

5. If a group of amateur astronomers, perhaps from the local astronomy club, has brought their telescopes to show the sky to children, make sure the children are aware that these people are donating their time and their telescopes and should be treated with respect. They may have even have built their telescopes themselves and are especially sensitive to unauthorized handling.

Introducing the Group to the Sky

Before anyone even thinks about looking through a telescope, get the group oriented to the night sky. If it is still dusk, then point out that the Sun has set in the west, and that the shadow of the Earth is rising in the east. At high altitudes, or in certain weather conditions, you might actually see the Earth's shadow and imagine it extending out into space. It is because we are in the shadow of the Earth that we can see the stars at night.

Pointing out the night sky should begin with a grand sweep across the heavens. This is the one time when a single bright flashlight is permitted, for you should use it to sweep the sky. In the fall, the expanse of the beautiful constellations of Cassiopeia, Cepheus, Pegasus, Andromeda, and Perseus is an invitation to tell the mythological story (see chapter 3) that relates these constellations. On winter evenings, the Heavenly G consists of bright stars that cover about a third of the sky. These stars are keys to the major winter constellations, such as Orion, Gemini, Taurus, and Canis Major. The Heavenly G is described in chapter 2. In spring, Leo, the lion, dominates the sky, but just to the lion's south is a very long, faint string of stars called Hydra, the sea serpent. The constellation is so long that it takes five hours to rise! Riding atop the snake are Corvus, the crow, and Crater, the cup.

In summer and early fall, the beautiful Summer Triangle governs the evening sky. The bright stars Vega, Deneb, and Altair form this triangle, and their constellations—Lyra, the harp, Cygnus, the swan, and Aquila, the eagle—occupy a large portion of sky. If you are away from city lights, you might see the Milky

Way march its way through Cygnus and Aquila, heading down to Sagittarius. Far south from the sky over the United States are the constellations of Scorpius and Sagittarius, which are very beautiful. Scorpius looks like a scorpion, and Sagittarius resembles a teapot out of which the Milky Way spurts like steam.

At any time of the year, Polaris is always visible in the same spot. It is the North Star, and it appears in one place because the Earth's axis of rotation points toward it. In spring, summer, and early fall, point out how the Big Dipper is a key to finding the pole. Use the two stars at the end of the Dipper's bowl to point north to the pole. No matter what time of night or time of year, if you see the Dipper, those two stars will always point toward the pole. The Dipper's handle is a key as well; its three stars "arc to Arcturus" and then "speed to Spica."

Using the Telescope

The ideal time to have an evening stargazing party is when the Moon is between a thin crescent and first quarter. As we noted before, these are the nights that the Moon is at its best, with its craters and mountains standing out in stark relief along the sunrise terminator. If you have selected one of these nights, you should concentrate your telescope view on the Moon, giving each child sufficient time to look, learn, and understand. Chapter 8 describes the Moon's highlights for each night of its orbital cycle.

If there are bright planets in the sky, the children should see them. Jupiter, king of the planets, is always worth a look if it is above the horizon. Its four moons will move relative to the planet from night to night, and on the planet itself you should see at least two dark belts, called the north and south equatorial belts. If Venus is in a crescent phase, it will look like a miniature Moon, and children enjoy hearing that this place is a world almost as large as ours, but one on which life as we know it cannot exist. Although Mars is the best known of Earth's neighbor planets, unless it is unusually close to Earth, it appears as a rather small disk.

Seeing stars is the next priority, and children often want to look at a bright star. However, a distant star, just a point of light to the unaided eye, is still just a point of light in a telescope. A telescope does not show more of a star, but it can show more stars. Where the naked eye can discern one star, a telescope might show two. Try looking at Mizar, the middle star in the Dipper's handle. In the field of view of the telescope, you will see four stars, of which Mizar, the brightest, is divided into two. Mizar is a binary star, and stars like it are well worth a view through a telescope.

Session's End

If children start to get tired or cold after a couple of hours, it is wise to wind things down, leaving them wanting more. Use your flashlight to check the grounds for candy wrappers, soda cans, and other trash. Turn on the security lights and reset sprinkler timers.

SUMMER CAMP NIGHTS

Few events are more splendid than a group of excited children looking up at a dark country sky from a summer camp. The country is where you get the best sky, and in summer the Milky Way stretches high across the sky. Also, summer is the time for meteor showers; the Perseids, one of the strongest showers of the year, is strong from August 8 to 12. With help from the Delta Aquarid shower, which peaks on July 29, the first two weeks of August are ideal for meteor watching.

First Night Out at Camp

At summer camp you will probably have the children for more than one night. If that is the fortunate case, we suggest using the first night to help them get their bearings and keep telescope viewing to a minimum. Include an overview of the basic aspects of the sky. Point out the Big Dipper and how it points to the pole, to

Arcturus, and to Spica. The Summer Triangle is also well placed and easy to point out. The Milky Way is high in the sky, and the teapot of Sagittarius is high in the south. It is also possible to tell the story of Andromeda, but only Cepheus and Cassiopeia are well up. Pegasus is low in the East and Andromeda and Perseus are not yet above the horizon as evening starts.

Second Night

Now is the time to bring out the telescope. Begin with the Moon and planets, and then go on to double stars. If the sky is dark, and the Moon is not interfering with that darkness, you might also try some deep sky objects (see chapter 15). The best choice for viewing on these summery nights is Messier 13, the great globular star cluster in Hercules. A true delight, M13 consists of some 100,000 stars packed so closely together that they resemble, as on 8-year-old claimed in amazement, "a little fuzzy poodle." This cluster is more than 20,000 light-years away; its light took that long to reach us. In a sense we are looking at this cluster as it appeared more than 20,000 years ago.

By this time the children are becoming familiar with the sky, and you are getting to know the children. As their knowledge grows their questions will become more probing and sophisticated; there is nothing like a dark sky and a distant cluster to inspire children to place themselves in the context of space and time. Don't be surprised if by the end of the second night you have a group of committed, thoughtful, and happy young skywatchers.

Third Night

It is a good idea to begin each night with a review of the sky and its major star patterns and constellations, and then further expand their knowledge by pointing out other fainter constellations. As the sky becomes richer to the children, they will soon start going off on their own, wanting to see more and more difficult objects.

If you and the children are up late, the constellation Andromeda will rise high enough so that the Andromeda Galaxy, one of the finest in the sky, will become visible.

Challenge Nights

In the constellation of Cygnus is an interesting planetary nebula called NGC 6826, which consists of a star surrounded by an outer shell of gas that it blew off some time ago. This object is informally known as the blinking planetary nebula. With a sizable telescope and magnification, try concentrating on the star in the center of the nebula. Under the right conditions, when you concentrate on the star the nebula should momentarily disappear! Then concentrate on the area around the star, and the nebula returns. You can actually make the nebula blink on and off.

On dark nights, try observing some distant galaxies. Messier 101 in Ursa Major is a large galaxy not far from the Dipper's handle. On a dark night through a good telescope, finding this galaxy is a challenge. As the nights follow one another, your children should find the sky a friendlier place, but one that gets more and more interesting with each passing hour.

Special Events

Aurora Borealis

The Northern Lights are a staple of observing in the northern part of the United States and all of Canada. They occur when magnetic energy from the Sun interacts with the Earth's magnetic field. They happen more commonly in Canada and the northern United States than they do in more southerly parts of North America. The frequency of displays depends on the status of the sunspot cycle, which rises and falls over an 11-year period (see chapter 7). At any stage but the minimum, there is a reasonable chance that on one summer night you will see the Northern

Lights. When the cycle is near maximum, the chances for displays are much greater.

Meteor Showers

It is at summer camp or during family vacations in national parks and seashores that meteor showers can be watched seriously. In chapters 10 and 11 we offer a program for meteor observation that, while difficult for schoolbound children with limited observing opportunities, might be ideal, if challenging, for summer camps where the opportunities for many observing nights are much richer.

Lunar Eclipses

Occasionally an eclipse of the Moon will occur during the summer months. If the eclipse is penumbral, in that only the Earth's outer shadow covers the Moon, you might detect only a slight shading. However, the shading allows the Moon's great ray systems, especially those of Tycho and Copernicus, to appear highlighted (see chapter 8). It is a little known fact that a penumbral eclipse is the best time to look at the Moon's ray systems.

If the eclipse is partial, meaning that the Earth's umbra or central shadow covers a portion of the Moon, then you will see the shadow gradually cover a portion of the Moon and later release it. In a total eclipse the shadow covers the Moon entirely, although the portion of the Moon closest to the shadow's edge will appear brighter.

The Sun

In chapter 7 we emphasize that one should never look directly at the Sun through a telescope. At summer camp the Sun offers a pageant of daily entertainment if there are sunspots. See-

28. Daytime observing. David Levy visits Rosina Stevenson's sixth grade, Pistor Middle School, Tucson.

ing the spots change in size and shape as they march over the Sun day by day—they take a bit less than two weeks to cross—provides a movie of the Sun as an active, changing star. If there is a partial eclipse during the summer, everyone will want to observe the Moon crossing the surface, blocking out sunspots as it goes. *See chapter 7 for safety warnings about the Sun!*

OBSERVING AT HOME

All of the suggestions in this chapter apply as well to a skywatching program held at home. However, the intimacy of a family skywatching session can provide an added dimension of fun and learning. Family activities need not be restricted to soccer games and ballet recitals.

A Family Discussion before the First Night

Before going outdoors, try engaging in a family discussion about the sky. We suggest emphasizing what objects and constellations will be best placed for viewing. If the Moon is visible, or if a planet will be well placed in the sky, emphasize the idea of the Earth's place in the solar neighborhood (chapter 1). When you have a chance to show children the Moon, emphasize the major features that they will be seeing (chapter 8). If the Moon is less than first quarter, discuss the effect of earthshine, and tell them about prominent features such as the "woman in the Moon."

First Night Out: Setting Up a New Telescope

If this is the first time you are using your telescope, we recommend that you spend the first night getting used to it. Read its instructions, set up the mounting, and attach the telescope so that it is properly balanced. Aligning the finder is a tricky but important part of this process. It really involves doing things backward. Place the lowest possible power eyepiece (the one with the longest focal length) into the telescope. Find a distant street lamp or light and focus the eyepiece on that light without using the finder.

The finder should be roughly centered in its bracket so that the three or six adjustment bolts are holding it at equal distances. With the telescope locked in position, adjust one set of three bolts until the lamp is centered in the finder. Then make sure that the lamp is still centered in the main telescope.

Now use the finder to locate Polaris, the north star. If the lamp is far enough away, Polaris should be centered also in the main telescope; if it isn't, then make fine adjustments to the finder bolts. The reason we use Polaris is that other stars will not stay centered in the telescope because the Earth's rotation moves them out of the field.

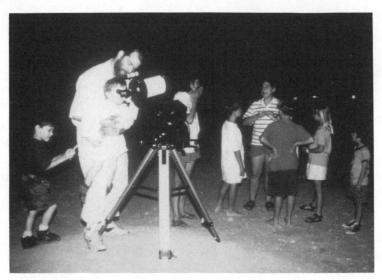

29. *Neighborhood Observing Night, 1995. Setup by Miranda Lebofsky, then 13 years old, in her backyard.*

We suggest spending some time getting familiar with the telescope before inviting family and friends to share it.

First Night Out once the Telescope Is Ready

A family skywatching evening session should be less formal than the school or camp sessions described earlier. Begin by setting up the telescope *in advance* of the start of observing. Introduce the sky in large gestures; the Big Dipper and pole star in the North, the Summer Triangle, or the Heavenly G in winter. Perhaps a few obvious constellations, such as Orion, could be included in this introductory romp about the sky. This part needn't last more than 15 or 20 minutes. Then, set the telescope up on the Moon or a planet, using low power first and then using higher magnifications.

Second Night

Getting familiar with both the sky and the telescope should be an enjoyable process that takes several nights. On this evening, try finding again the objects you looked at the first night. The Moon's changing phase will mean that different craters stand out, and Jupiter's moons will appear in different positions relative to the planet. You should also notice that the telescope is becoming easier to use as you get more familiar with it.

Third Night

From a basic knowledge of the sky's main features, you now can point out several specific constellations, such as Leo in spring; Lyra, Cygnus, Scorpius, and Sagittarius in summer; Pegasus and Andromeda in autumn; and Auriga, Taurus, Orion, Canis Major, and Gemini in winter. With increasing ease, you are able to enjoy

30. Observing Night. Agua Caliente Elementary. Spring 1996.

the telescope more. Instead of concentrating on the Moon and planets, you now can move on to some of the sky's other treasures. Try the double star Mizar (spring); the double Beta Cygni or the globular star cluster Messier 13 (summer); the Andromeda Galaxy in autumn; and the Great Nebula in Orion (winter).

Following Nights

Never give up. After spending a night or two under the stars, many families put their telescope into a closet and don't use it again, preferring instead the indoor comfort of the television set. Going outdoors, especially on a cold evening, is more of a challenge, as is being bitten by mosquitoes in summer. However, the evenings need not be long, and quiet nights under the stars can be very pleasant indeed. The goal is to show children that the sky is a part of our natural world. As much as the plants, animals, and other aspects of the planet we live on, the sky is a fascinating place that children can enjoy.

JOINING AN AMATEUR ASTRONOMY CLUB

As your interest grows, you might want to join an astronomy club. There are local societies in almost every city in North America. They typically meet twice a month, once for a regular meeting, and once for a star party or observing session. Some clubs have junior or children's sections that meet before or during the main meeting.

Club star parties can be especially helpful if you are considering buying your own telescope. There might be several dozen telescopes set up, including small ones that you are considering. If the star party is at a remote site, the club newsletter usually provides directions to get there. Remember that clubs do operate on a volunteer basis; although most members are happy to assist newcomers, you will probably find that some are more helpful than others.

Major Annual Conventions

Across the United States and Canada, amateur astronomers get together from time to time to celebrate their hobby and a few good nights of observing. These events can be truly spectacular; it is not unusual for telescopes as large as 30 inches in diameter or more to appear like leviathans. Meals and lodging are usually available, and often there is a program of presentations during the day. Dates and contact numbers are provided in the *Sky and Telescope* and *Astronomy* magazines. A sampling of the better known conventions follows:

Stellafane: Started in 1926 by Russell Porter on a hilltop near Springfield, Vermont, this meeting, which attracts some 3,000 people each summer, is by far the oldest. It is traditionally held on the weekend of the "first dark of the Moon after the Green Mountain corn is ripe," in late July or early August.

Riverside: Founded in 1969 by Clifford W. Holmes, this conference is held each year on the Memorial Day weekend at the end of May at a YMCA camp near Big Bear Lake, California. It draws upward of 1,500 people from California, other western states, and even western Canada. *Nightfall*, held each September in southern California, is a small spinoff from this event.

Texas Star Party: Held under the pristine skies of the Davis Mountains in west Texas, this star party offers a week of observing with large telescopes each spring.

Winter Star Party: Each year, usually in February, the Southern Cross Astronomical Society sponsors this event, which is held near Big Pine Key in the Florida Keys. Be aware, however, that the limited space of this facility means that the attendee list usually fills up several months before the event takes place.

Astrofest: Held each September by the Chicago Astronomical Society, this meeting is one of the largest in the midwest United States.

Hidden Hollow: Ohio's Richmond Astronomical Society sponsors this star party every two years, usually in late September or early October. In addition to offering many small telescopes, Hidden Hollow features a giant 31-inch equatorially mounted telescope and observatory.

Starfest: Sponsored each summer by Canada's North York Astronomical Society near Toronto, this event is Canada's largest.

Mount Kobau: In Canada's Rocky Mountains each August, this star party takes place atop a mountain that the Canadian government once reserved for its Queen Elizabeth II telescope, a 150-inch-diameter giant. Although the telescope was never built, the site still serves for this annual star party.

MAGAZINES AND ONLINE SERVICES

The two major astronomical magazines are *Sky and Telescope* and *Astronomy*. These magazines appear monthly and are available at newsstands or by subscription. They contain articles and listings of coming events in the sky. Although *Astronomy* is aimed at a more introductory market, *Sky and Telescope*'s articles are also well written and easy to follow. *Odyssey* is a magazine specifically aimed at elementary school children. In Canada, *Sky News* is aimed at the beginning skywatcher.

Most of the major online services offer astronomy areas. CompuServe has an astronomy forum that is set up with different interest areas ranging from a beginner's section to a CCD corner for observers using electronic systems. The advanced members of the forum are usually very helpful in answering even the most basic questions such as "What was that bright object I saw in the west last night?"

The Internet offers several astronomy-related newsgroups led by sci.astro. Because these groups tend to be unmoderated, their message threads can be rather strange at times. We suggest a liberal use of your program's ability to eliminate unwanted contributions and contributors. Theories that claim that the Universe doesn't operate on known physical laws are typical of the kind of postings that can confuse the uninitiated.

SPECIAL PROGRAMS

Bringing specialists into the classroom is a grand idea. It gives the children a new perspective and helps to bring together teachers, children, and astronomers. We find that the best use of a professional astronomer is to learn why he or she became an astronomer. Children want to know how an astronomer spends a day and a night and what lies behind the astronomer's passion for the sky.

One of the best groups ever to appear in a classroom is *Astronomy to Go*. Run by amateur astronomers Bob and Lisa Summerfield, the program is privately sponsored through the sale of

31. *Thea Cañizo and her telescope, at open house, Vail Middle School. Spring 1996.*

T-shirts and other astronomical paraphernalia at star parties and science teachers conventions. Besides the Summerfields' enthusiasm, the program's highlights include large telescopes and meteorites. The whole point of teaching astronomy, Bob Summerfield thinks, is to "wow" people. Imagine a class of third-grade children watching the setting up of a 20-inch-diameter telescope. "That's for us?" their open mouths would sputter.

"Dead aliens" is one of *Astronomy to Go*'s favorite programs. The Summerfields bring their collection of alien meteorites into the classroom and get the children to hold them and consider their long journeys through space. *Astronomy to Go* has made presentations in classrooms across the United States. Even though they are respected amateur astronomers, the Summerfields would forego a private evening with his telescope if a school group needed their services. "Sharing the sky with the public," Bob Summerfield insists, "is what our lives are all about."

How to Turn Kids Off to Astronomy: A Top Ten List

10. *Emphasize that astronomy is difficult, mathematical, and has no relation to everyday life or culture.*
9. *Use a telescope with a wobbly mount and inferior optics.*
8. *During a solar eclipse, lock the children up indoors with the curtains drawn.*
7. *Include astronomy as a brief segment inside a geography unit. This way the children will see the Universe as a minor offshoot of the Earth.*
6. *Rush children through the line at the telescope.*
5. *Avoid interactive, "hands-on" activities that are designed to get the children to learn by doing.*
4. *Don't plan skywatching parties well in advance, or set them up in well-lit fields!*
3. *Do not begin by having them look at the Moon. Choose an esoteric, difficult-to-see object.*
2. *Teach the sky as a subject independent of all other subjects.*
1. *Keep the fun out of it!*

Part III

Our Neighborhood in Space

Chapter 7

Safe Sun

Respect the sun. Do not look at the sun without proper protection for your eyes. A single unprotected look through binoculars or a small telescope can cause permanent eye damage.

Most people think of the starry sky as a place to be studied and appreciated only at night. They forget that the biggest, brightest star of all rules the sky by day. The Sun is a fascinating place, busy and churning. It would be the ideal astronomical object for children, if it were not for the fact that a single look through an unprotected eyepiece will blind a child forever.

Therefore, we begin our chapter with a warning: **Do not look at the Sun without proper protection.** *A prolonged look at the Sun with the unaided eye can cause damage. The problem is that it is important to stress this danger to all the children without lessening their interest in the Sun. To accomplish this, we suggest a two-pronged approach to solar observing. The first part is intended to emphasize the danger, and the second is to show how we can study and enjoy our star.*

WHY IT IS DANGEROUS TO LOOK AT THE SUN

We do not see the full spectrum of rays that bombard us from the Sun. Suntans and burning on a hot sunny day come from ultraviolet rays that we cannot see. It is also easy to explain that when solar rays are focused, they become far more intense and therefore more dangerous.

DEMONSTRATION: SAFE SUN, PART I

PURPOSE: To show the children that even a brief look through a telescope can be dangerous.

MATERIALS:

A small telescope, either a Newtonian reflector or a refractor. If your telescope mirror is 6 inches in diameter or larger, consider stopping it down to reduce the light that gets reflected by the mirror to the eyepiece. Many small telescopes have dust caps with a smaller plug that can be removed. You can leave the dust cover in place and just remove the plug. If your telescope does not have this feature, we recommend cutting out a piece of cardboard the size of your telescope's front end. Cut a circular hole in the cardboard. If your telescope is a reflector, offset the hole from the center so that the telescope's secondary mirror does not block out too much light. Then tape the cardboard to the front end of the telescope. The purpose of this device is to reduce the amount of light and heat coming through the telescope.

A sheet of thin paper or a plastic garbage bag, but all you need is a piece about a foot square. Do not use the fireproof bags that some companies produce. They might not work, and they bias the experiment; after all, your eye is definitely not burnproof!

An additional teacher or parent to make sure the children do not look through the telescope.

Procedure:

1. Focus the telescope. Before using the telescope on the Sun, focus it on the most distant object you can find, such as a far-off mountain or tree.
2. We recommend a brief discussion about the strength of the Sun's rays. Mention squinting, the body's natural defense against too much sunlight—if the children look up at the Sun, their eyelids start to close, in a wonderful and natural protective measure. This automatic reaction, however, is not nearly sufficient when using a telescope, which concentrates the rays of light.
3. Make sure the children are sitting in a group on the side of the telescope opposite from the eyepiece.
4. Hold up the plastic trash bag with one hand gently and punch it with the other fist. "Is your eye as strong as this?" you ask them, encouraging them to imagine the pain if someone hit their eye that hard. Once they understand the concept that their eyes are more delicate than the material in the garbage bag that is to be used in the experiment, you are ready to proceed.
5. Protect the telescope's finder. If the telescope has a small finderscope attached, make sure that the lens facing the sky is covered to prevent it from being directed at the Sun.
6. Set up the telescope. In a sense, the Sun finds itself with your telescope, since the scope casts a shadow. As you grasp the telescope, look at the shadow that it casts. The idea is to move the telescope tube so that its shadow is as small as possible. When that happens, you should see bright sunlight streaming through the eyepiece. *Keep your eye away from the eyepiece.*
7. Make sure that the children are not near the telescope.
8. Tell the children that the first part of this demonstration is intended to show the power of the Sun when viewed through a telescope and its ability to cause instant blindness. Hold up the piece of garbage bag and say that instead of placing an eye

at the eyepiece, we will place this plastic bag there. Ask them how long they think it will take for the Sun's heat to burn a hole in the bag.

9. Holding the garbage bag with two hands, spread it out about 6 inches away from the eyepiece so that the Sun projects a circular image on the bag. Slowly bring the bag closer to the eyepiece until the Sun's image is a sharply focused point of light. At or before this moment, the bag should ignite. The burning should be limited to the small area around the bright spot of sunlight and should stop when you take the bag away. N.B. *Make sure there are no flammable materials near the telescope. We do recommend, as a precaution, that you have a glass of water or fire extinguisher nearby.*

10. As soon as the experiment is completed, turn the telescope tube away from the Sun. Prolonged sunlight could actually break the eyepiece.

DISCUSSION: No doubt now: the Sun's rays are harmful! Restate how dangerous the Sun can be to your eye, and just as we cannot repair the damage to the bag, we could not repair the damage to the eye. One precocious 8-year-old asked us if the burning plastic bag would harm the environment. Our experience is that the burning is limited to the few seconds when the bag is actually in contact with the focused rays of the Sun and that the instant you remove the plastic the burning stops. We think that the point of the demonstration—to emphasize the danger of looking directly at the Sun through a telescope—is worth it. The problem with using plain paper is that the burning sometimes does not stop after you remove it from the source of heat, so have some water handy.

TRANSITION: If it is not safe to look at the Sun, how are we going to study it? One answer is to do it by projection, the same way that movies are shown. The telescope becomes the movie projector, and we don't need a bulb since the Sun provides its own light. We are now ready to enter the part of the demonstration in which we actually see what lies on the surface of our Sun.

Demonstration: Safe Sun, Part II

PURPOSE: To observe the Sun safely and comfortably.

MATERIALS:

Small telescope

Additional teacher or parent to make sure the children do not look through the telescope

A piece of white cardboard, at least 2 square feet, or a small box with white paper at the bottom

PROCEDURE: Now that we have completed our demonstration about the need to look at the Sun in safety, we will set up the telescope in such a way that we can study the Sun in complete safety by looking at the projected image. Emphasize again that the children are to look at the projection and not through the tele-

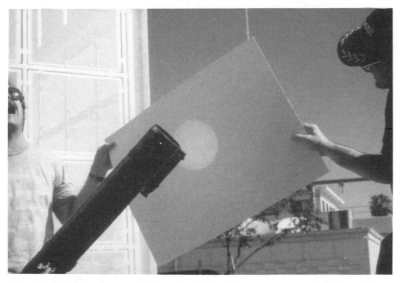

32. *Jonathan Becker and Larry Lebofksy demonstrate a Safe Sun observation.*

33. *Safely viewing the Sun.*

scope, just as at a movie theater they would watch the screen and not look into the bulb of the projector. Add that this method has the two advantages that it is safe and can be done quickly. Instead of the children lining up to look one by one, they can all see the projected Sun at the same time.

1. Check to make sure that the children are sitting opposite to where the eyepiece is pointing and that the finderscope is covered.
2. Find the Sun as before, by using the telescope's shadow and making it as small as possible.
3. Use the lowest-power eyepiece you have, the one with the longest focal length engraved on its barrel. When you see sunlight pouring through the eyepiece, place the white cardboard a foot or two away. The Sun should form a circular image.
4. Focus the telescope so that the edge of the Sun appears as sharp as possible.

5. Make sure that the Sun's image appears as a full circle. If any part of the image is cut off, move the telescope tube slowly until you see a full circular image. If it is impossible to see the full image, then you are using too high a magnification.
6. This is the actual observation. You and the children are actually looking at a star in the sky. There may be dark areas called sunspots. Jiggle the telescope a little; as the Sun's image moves, the spots should move along with it. (If they do not, then you are probably just seeing dirt on the eyepiece.)

The Sun doesn't always follow our schedule, and there may not be any spots on the Sun the day you decide to observe it. This is especially likely if the Sun is near a low point in its 11-year cycle. But if you do see a spot at all (which is probable starting in 1997) point out to the children that it is quite large, perhaps a significant fraction of the size of our whole Earth. We are now looking at a storm on the Sun: not a rain storm, but a magnetic storm. The Sun is cooler where the spots are. Sometimes these spots are accompanied by slightly whiter spots called faculae.

The observation does not have to last more than 5 or 10 minutes—just long enough for everyone in the group to see the Sun and any visible sunspots.

DISCUSSION AND CLOSURE: Is the Sun active today? How many spots did we see? Were the spots spread out evenly across the Sun? Did some appear to be clustered in groups?

The discussion could end by emphasizing the value of the Sun. Point out that the children have just had their first scientific look at a very special star that provides the heat and light needed to sustain life on Earth. If the star we just looked at were to be blocked out for several months, plants would not be able to conduct photosynthesis, the process by which sunlight is converted into chlorophyll, and we would all eventually starve. The temperature on the planet would also drop.

Recording the Positions of the Sunspots. Ask the children to record the locations of the spots they see on the Sun. The beauty of projecting the Sun in this way is that the whole exercise need not last as long as it would if the children had to line up one at a time to look through the telescope. The class could have another look at the Sun the following day and see how far the spots have moved across the Sun.

Ultimately, this process should show students that when looked at safely, the Sun is a place that changes from day to day. Observing the Sun will give the children a personal acquaintance with a star.

Repeating the Demonstration. With a small telescope and easy access to the outside, it is possible to repeat the second part of *Safe Sun* every sunny day for a week or two. This way, your class will see the progress of sunspots as they march across the rotating Sun; it takes some two weeks for a spot to cross from one side of the Sun to the other.

Drawing the Sun. If you launch an ambitious daily sunspot-watching project, encourage your group to draw what they see. A simple circle traced on a piece of paper, using a plate or a cup as a guide, will serve to outline the Sun. Have the children draw the positions of the spots on the paper as accurately as they can, and ask them to include the date, time, and number of spots seen. This way they will have a record of the changing positions of the spots as the Sun rotates.

Using a Filter

An alternative to projecting the Sun's image is to use a filter. The only type of filter currently in production that is safe enough to use is the one that fits over the front end of the telescope tube. Such a filter blocks out almost all the Sun's rays before they get concentrated by the telescope's optics.

WARNING: If you bought a small telescope equipped with a small solar filter that attaches to the eyepiece, throw the filter away. It works right at the focal point, where the Sun's rays are hottest and thus can easily break, giving you an eye full of blinding sunlight!

Front-end filters can be made either of glass or of an aluminized Mylar plastic. With one of these, you and the children can look through the telescope's eyepiece, seeing the Sun directly and safely. The advantage is that as long as you make sure the filter stays on, there is no chance that a child will accidently look through an unprotected eyepiece. However, children do play with things. We do not recommend this approach unless you can absolutely guarantee that the children will not be able to remove the filter. The other disadvantage is that the process takes much longer, since every child has a separate look.

Sight the Sun the same way as before, by finding where the shadow of the telescope tube is smallest. The Sun should appear as a disk of light through the telescope, and it should be completely comfortable to view. If the image causes eyestrain, then the filter is not strong enough and should be discarded.

Make sure that each child gets a good look. You should check the eyepiece every 2 minutes or so to make sure that the Sun is still in the field; unless the telescope has a motor drive to counteract the Earth's rotation, the Sun will need to be repositioned often.

THE SUN AS A STAR

Surprisingly, some people have suggested that the concept of the Sun as a star is not appropriate for elementary school children. We have found that young children readily grasp the idea if we show that the Sun, like the Moon and planets, is a different type of place. "Our Sun is a star," we say, "and every star is a sun." The only reason the other stars are not as bright as our Sun, we add, is that they are far away.

Our Sun is the provider of all the heat and light that support life on Earth. Its light allows green plants to perform photosynthesis. Its heat drives the storms that bring rain. But despite this, we tend to take the Sun, this monstrous nuclear furnace where 4 million tons of hydrogen are fused to helium every second, quite for granted.

FACTS ABOUT THE SUN

By far the largest object in our solar system, the Sun is about 860,000 miles, or 1.4 million kilometers, in diameter. It shines through nuclear fusion that takes place at its very center—the *solar core*—where hydrogen fuses to helium. The solar energy then bubbles upward toward the surface through a large area called the *convective zone*. After some time this energy reaches the Sun's visible surface, which we call the *photosphere*. This area is only about 400 kilometers thick. If our telescope is large enough and the sky is free of turbulence, we can see the tops of the convective bubbles as a pattern of *granulation*. The *sunspots* are caused when turbulence below the surface interacts with magnetic fields. Sometimes these sunspots are accompanied by bright regions called faculae.

The Sun's surface temperature is about that of an iron welding arc, some 11,000 degrees Fahrenheit, or 6,000 degrees Centigrade. The sunspots are almost 2,000 degrees Centigrade cooler than the surrounding surface.

Through a special filter designed to accept only the Sun's hydrogen-alpha light, we can see other features on the Sun. *Flares* are the result of the great magnetic forces in the solar atmosphere that are released explosively from time to time. *Prominences* can rocket tens of thousands of kilometers about the solar surface, and come from lines of force that connect magnetic fields around sunspots.

The Sun's atmosphere consists of a tenuous layer called the *chromosphere* and a large area called to *corona*. We need a total eclipse of the Sun to observe the full extent of the corona, whose temperature climbs into the millions of degrees.

Much of the Sun is rarefied. The corona has less material than the best vacuum we can create on Earth, and the photosphere, though very bright, is about as thin as the space above the Earth where the shuttle flies. We would have to move a tenth of the way from the photosphere to the core before we found material as dense as the air in our own atmosphere.

SUNSPOTS

More than 2,000 years ago, Chinese observers noted some dark features on the Sun. But it was Galileo who pointed his telescope toward the Sun and recognized that these features were events of some sort that took place on the Sun's surface. It is possible, by the way, that Galileo paid for his discovery with his sight; in later years he was almost blind.

We now know that the sunspots are great magnetic storms that appear dark only because they are cooler than the rest of the surface. We also know that these storms come in 11-year cycles. If you conduct Safe Sun when the sunspot cycle is at its minimum, it is quite likely that you will have no spots to see at all!

The discovery of the sunspot cycle dates back to 1826 and the small town of Dessau, Germany. In that year an amateur astronomer named Heinrich Schwabe bought a small refractor for himself. After doing some stargazing he wrote to an astronomer for advice on what useful work he could do. In his reply, the scientist advised that he study the Sun to search for Vulcan, a hypothetical planet so close to the Sun that it might be found only as a black speck when it crossed directly between Sun and Earth.

Schwabe began this experiment, observing the Sun every clear day and noting the positions of the sunspots. After 17 years, he had not found Vulcan, but he did announce that his own observations had shown that the sunspots appear in a cycle that lasts about a decade, which is pretty close to the 11-year period we know of today. For this discovery, the skywatcher was awarded the highest astronomical honor in England, the Gold Medal of the Royal Astronomical Society. By counting the sunspots over a period of time, we can recreate Schwabe's discovery.

ECLIPSES OF THE SUN

If the most inspiring sights could be listed as the seven wonders of the world, a total eclipse of the Sun would be on top. Seeing the Moon get exactly between the Sun and Earth and slowly blot out the Sun is a thrilling sight. When the Moon first begins to cross the Sun in the sky, it looks as though something is taking a bite out of the Sun; no wonder ancient people thought a dragon was gobbling up our source of light and heat. As the Moon continues moving eastward, the Sun becomes crescent shaped. After the midpoint of the eclipse, the Sun's full shape and light slowly return.

If you are lucky enough to be in the narrow path of a total eclipse, the crescent of Sun gets smaller and smaller until only a sliver is left. From the west (or northwest or southwest, depending on where the path of totality is located) a dark shadow rushes toward you. Suddenly you are enveloped in darkness; you look up at the Sun: it is gone. In its place is a jeweled crown.

34. *Solar eclipse. Courtesy Flandrau Science Center.*

Usually the total phases of solar eclipses last no more than 2 or 3 minutes. However, in July 1991 the Moon, almost as close to the Earth as it can be, covered the Sun for almost 7 minutes, and for those of us who witnessed this event, it was an incredible period. It looked as though Nature itself was shutting down during the spectacle—animals were quiet, the breeze died, and in place of the Sun was this strange circle of darkness surrounded by two ears of pearly light.

A Ring of Fire

If the Moon is near the apogee, or farthest point, of its orbit around the Earth, it will not completely cover the Sun. During the maximum phase of such an eclipse, the Moon is surrounded by an annulus or ring of sunlight. On May 10, 1994, such as eclipse was visible in cities across the United States. One such city was Las

35. *Safely viewing the Sun.*

Cruces, New Mexico, where Wendee Wallach included observing the eclipse as part of her physical education class at Sierra Middle School. Under her careful supervision, children were allowed to observe the eclipse wearing specially made glasses.

Observing Eclipses

*WARNING: **Never look at a partially eclipsed Sun without proper eye protection.** A welder's glass, #14 strength will protect you from the Sun's rays. The danger of blindness is greater during eclipses because the low light level allows us to stare at the Sun for long periods without squinting. During the moments of totality, absolutely no sunlight gets through; it is safe to look at the total phase of an eclipse. Several companies offer special aluminized Mylar "glasses" that should be safe for eclipse viewing. When using these, make sure there are no holes in the Mylar. The onset of totality is marked by a sudden drop in surrounding light; in a clear sky you can tell when totality has begun when you cannot see any sliver of Sun through the glasses.*

Whether an eclipse is partial or total, as the Moon covers more and more of the Sun, it will obscure sunspots. Observing a partial eclipse is similar to observing the Sun in ordinary times. During totality, however, things get more interesting. This is the only time on Earth we can see the Sun's hot and tenuous outer atmosphere or corona. We can also see prominences, the leaps of gas from the surface of the Sun.

Solar Eclipses

Date	Type	Approximate area covered
February 26, 1998	Total	Pacific, Central America, Caribbean
August 21–22, 1998	Annular	Malaysia, Indonesia, Philippines
February 16, 1999	Annular	Australia
August 11, 1999	Total	Atlantic, England, Europe, India
February 5, 2000	Partial	Antarctica

WE CAN'T LIVE WITHOUT THE SUN

The very end of a Safe Sun demonstration, whether it lasts an hour or two weeks, should involve a discussion of what would happen if we didn't have the Sun. It is possible that after a large comet or asteroid hit the Earth 65 million years ago, a dense cloud blocked out the Sun worldwide for many months. With photosynthesis stopped for that long, worldwide starvation extinguished more than 70 percent of all the species of life. Reassure the children that such an impact is very rare.

Many children ask, "When will the Sun stop shining." It has been shining at its current level for several billion years and is expected to continue that way for some 5 billion years more. In chapters 13 and 14 we will look at other suns that are not so stable—suns that change radically in brightness over periods of months and suns that can brighten by a hundredfold in a single night. We are very lucky to be under the watchful eye of a star as stable as the Sun.

ACTIVITY: STAINED-GLASS SUN SYMBOLS

BACKGROUND: Stories about and artistic representations of the Sun, Moon, and stars can be found in many cultural traditions worldwide. Sharing these different points of view allows children to see a value in traditions other than the Man in the Moon or the Sun as a happy face. Once students have basic factual information about the Sun, Moon, and stars, introducing them to nonscientific interpretations by observers in other times and places allows them to enjoy different perspectives.

OBJECTIVES: Students will explore the different artistic views of the Sun as expressed by ancient and contemporary cultures. They will create a Sun symbol: a Sun-catcher based on a traditional design. Options: Students can create a construction paper or rock art symbol of their own design.

36. *Sun symbols. Left–right: Ana-Alicia Robles, Elizabeth Redondo, Thea Cañizo.*

MATERIALS:

Cardboard backing, one per student

A heavy-duty plastic wrap or transparencies

Permanent markers

Masking tape

Cellophane tape

Pictures of Sun symbols

PROCEDURE:

1. Present grade-appropriate factual information about the Sun.
2. Introduce the concept of a picture or a symbol representing an object. Discuss Sun symbols; e.g., differences, similarities, importance to the people who designed them, scientific accuracy (or lack thereof). Use Sun symbols, as well as those found in clothing and jewelry catalogs, calendars, etc.
3. Share Sun legends, myths, and folklore from a variety of countries with the students.
4. Construct stained-glass Sun symbols:

   a. Ask each student to choose a Sun symbol. Carefully tape the picture by the edges to the work surface with masking tape.

b. Provide scrap cardboard cut into rectangles slightly larger than the Sun symbols for backing. Paper plates also work for round Sun-catchers.

   c. Tear off a sheet of plastic wrap several inches larger than the cardboard backing. Place the plastic wrap over the Sun symbol (a heavy wrap is less likely to tear or leak).

   d. Trace the symbol with permanent markers. Bright colors, black outlines, and filled-in areas work better than light colors and fine-line outlining. *Note*: Transparencies work better than plastic wrap, but are expensive.

   e. Tear a piece of aluminum foil a little larger than the cardboard backing.

   f. Crumple the foil *gently*, then carefully flatten it. The uneven surface will catch the light, creating a stained-glass effect.

   g. Lay the foil over the cardboard, folding overlapping edges to the back.

   h. Place the plastic wrap over the foil, folding overlapping edges to the back. Tape securely with cellophane tape.

   i. As an alternative, cut out the centers of two plastic plates, sandwich the plastic wrap between them, staple the edges of the plates together, and trim away any excess plastic wrap. This creates a see-through effect. Plastic wrap that stretches easily works best for this option.

5. Option: Ask the students to design construction paper Sun symbols of their own. Discuss the significance of each completed symbol. Use the symbols for a bulletin board display.

6. Option: After they have created a Sun symbol on paper, students can use paint or markers to transfer their symbols onto smooth rocks.

7. Option: Cut brown paper grocery bags into squares. Ask the students to crumble the paper until it is soft and wrinkled. Using water colors or markers, have the students put their Sun symbols on the brown paper to simulate rock or cave paintings.

8. Allow the students to write their own legends. Suggested topics: how the Sun got into the sky, sunspots, solar eclipses, or why we don't see the Sun at night.

Chapter 8

Enjoying the Moon

Besides being the brightest and easiest thing to see in the night sky, the Moon is the most important object to observe, especially for children. The Moon is so bright that a teacher or parent can actually see the cone of its light entering the eye of the young observer as he or she looks through the eyepiece of a telescope and can thus make sure that the child is getting a good view.

Seeing the Moon through a telescope is the first step toward visualizing it as an actual place. A child can imagine hiking up and down the mountains, climbing into the craters, and examining the dark gray rocks. Then come the questions: how did all those craters get there? and if the Moon has craters, shouldn't the Earth? Because the Moon is close enough to be seen as a place, we devote more space in this book to the Moon than to any other object in the sky.

OBSERVING THE MOON FOR THE FIRST TIME: AN ACTIVITY

Because the Moon is bright, easy to find, and shows so much detail, it should be the first object that children view. Take your time with this important observing session and follow these steps:

MATERIALS:
Telescope
Chair or ladder to allow shorter children to reach the eyepiece
Moon map
Observing logs

PROCEDURE:

1. If schedules and weather permit, plan your observing session for a time when the Moon is between 4 and 8 days after its new phase—from its thin crescent phase to just after first quarter. This is the time when the Moon offers its most pleasant view— it is not too bright, and the dark portion is still visible. However, if such a time is not convenient, at least make sure that the Moon is in the evening sky—which it is between 3 and 16 days after new Moon.

2. It is vital that this first look be a relaxing, special experience. Even if the line of children waiting to view is long, do not rush them through it. At the beginning, tell the group that each child, even the one at the end of the line, will have as much viewing time as he or she needs.

3. Explain that the Moon is a place, with ancient craters and basins of great age. Ask them to look for features in this order.

   a. Light and dark regions: The dark regions are the plains of lava that were formed some 3.9 billion years ago. The bright regions, also called the lunar highlands, are much older.

   b. The craters: Explain that these craters are mostly the result of ancient impacts. When the Moon and Earth were young, they were bombarded with asteroids and comets that formed crater after crater. The Earth did not keep her early craters; thanks to the never-ending processes of mountain-building and erosion, they have vanished. But the record of the Moon, especially in its ancient highlands—the bright areas we see—goes back almost to its beginning. By looking at the Moon, we see our own past.

4. As the children look through the telescope at the Moon, make sure that the cone of moonlight actually falls on their eyes. As soon as that happens, you can expect smiles of recognition and awe as they see the old world, but so new to them. Without a telescope, the Moon is a bright gray light in the sky. But through a telescope, it becomes a place with craters and mountains.

CLOSURE: Ask the children to describe what they saw, with possible drawings, in their observing logs. Discussion should invoke the following material.

How Did the Moon Get There? Although almost all the planets have moons, ours is different. One of the largest moons in our solar system, it circles a relatively small planet: we can almost call the Earth and Moon a double planet. Only Pluto, with its moon Charon, is similar in that sense.

Several theories of the origin of the Moon have been proposed over the centuries, one of which says that the Moon was formed out of the cavity that is now the Pacific Ocean. However, even though there is a similarity in size, there is evidence that this theory is not correct.

During the 1980s a new theory was proposed. Shortly after the formation of the solar system, the idea goes, a large body, possibly as big as Mars, collided with Earth. It would have been a sight to see, had anyone been on Earth to see it—a big approaching object in the sky, many times brighter than the full Moon and getting brighter every second—and then there was a tremendous crash, the sound of the ages as the object smashed into Earth at a speed of thousands of miles an hour. The force of the collision melted the Earth's crust, and huge quantities of material from both Earth and the pulverized other body formed a large ring of material around the Earth. Over a long period of time, perhaps a few hundred million years, the pieces of the ring collided with one another, and eventually accreted to form what we now call the Moon.

Hundreds of millions of years later, comets and asteroids of many sizes were still plummeting into the Earth and the Moon. We have no record on Earth of these impacts, but the remains are visible on the Moon as impact basins. Some 3.9 billion years ago, the largest of these, called the Imbrium Basin, was carved out over a large part of the Moon's surface. After the passage of more hundreds of millions of years, these basins filled with lava that welled up from deep within the still-hot subsurface regions. The dark areas we see on the Moon are these lava-filled plains, which we call *maria* (or *mare* for singular). Although mare is a Latin word meaning sea, these plains do not contain so much as a puddle of water. They are, rather, made of a dark rock called basalt. (In 1994, the Clementine spacecraft studied the Moon's surface and recorded evidence of ice in the Moon's permanently frozen south polar region.)

Formed so long ago, the maria represent a lunar repaving project that tells us much about the Moon's history. Compared with the rest of the Moon's surface, there are not many craters in these areas. Thus we know that most of the Moon's craters were formed by impacts during the half-billion years between the Moon's formation and the creation of the great basins; another half-billion years elapsed before the basins filled up with lava. All but one of these plains are on the side of the Moon that faces us. Mare Orientale is the exception.

Over the years, there have been several theories of crater formation on the Moon. But research in the last 25 years, which includes walking and driving around these craters, has shown conclusively that virtually all the lunar craters are the results of impacts. Although most of these impacts took place during the early years of the Moon's existence, a time we call the *period of heavy bombardment*, some, such as Tycho, are much younger, formed perhaps 100 million years ago. Earth must have been subjected to this kind of pounding as well but as we noted before Earth generally does not keep her craters; the forces of time erode, remove, or cover them, and only a few remain. Some 50,000 years ago, an iron-rich asteroid 30 yards wide crashed into what was then a northern Arizona forest, creating the big crater we still see today near Flagstaff.

Looking at the pock-marked Moon, we can all get a sense of it as a place with a history, and, at the same time, it should help us to appreciate the long and violent past of our own planet.

Why Does the Moon Always Point the Same Side toward the Earth? The major part of the long history of the Moon is its connection with the Earth. This relationship is very much controlled by the mutual gravity between these two bodies—the tidal force between the two worlds. On Earth we see this in the ocean tides. Twice a day, all over the world, the oceans rise and fall. In mid-ocean the water rises and falls by about a foot, but the tidal force is magnified at the shorelines by varying amounts. The highest tides on Earth are found at the eastern end of the Bay of Fundy in Nova Scotia, where a volume of water equal to the combined flow of all the rivers on Earth fills and then empties the adjoining Minas Basin. The resulting tides in the nearby rivers and tributaries have been known to exceed 50 feet.

It is a little known fact that the Earth exerts a tidal pull on the Moon as well. Over the years this pull had the effect of slowing down the Moon's rotation, until it reached a state we call "rotational lock." The Moon's rotation period now equals its 28-day-revolution period about the Earth: as the Moon orbits the Earth it now always keeps the same face toward us. The tidal pull of the Moon on the Earth is gradually slowing down the Earth's rotation too, but so slowly that the Earth will not be rotationally locked with the Moon for billions of years.

THE PHASES OF THE MOON

Explaining the Moon's phases to children is difficult because it involves thinking in three-dimensional terms. The Moon goes through phases because it is traveling in a path that takes it around the Earth—which we call an orbit—once a month and as it moves around the Earth it presents a different angle to us each night and day relative to the Sun. If the Moon is between us and the Sun, then all of the Sun's light shines on the side we do not see. If it is a quarter of the way round, then we see half of its face illuminated. If it is opposite the Sun, then we see it fully illuminated.

When the Moon is near its crescent phase, it is possible to see its unlit side. If we were standing on the darkened part of the Moon facing Earth during this time, we would see the bright Earth lighting up the sky, and it is indeed the Earth that is lighting the dark part. We call this dim light earthshine—the effect of the Earth's shining on the Moon.

A Monthly Romp across the Moon

An ideal time to observe the Moon is when it is in its crescent phase. So little of the Earth-facing part of our satellite is illuminated by the Sun at this time that the Earth-lit dark side is easily visible. Through a telescope, the Moon's ghostly dark side shines nicely, allowing an observer to take make-believe walks across its mountains and craters.

We now look at the appearance of the Moon each night, from new Moon to full and back again. As a guide, the Moon reaches first quarter at about day 7 past new, full at day 14, and last quarter at day 21. Naturally some nights will likely be cloudy, but on those that are clear, you should be able to spot some of the features we describe. See Moon Map on page 151.

DAY 1: Although a few stalwart observers have seen a Moon younger than a day, it is rare to see the thin crescent Moon this early after new phase. There is a hard-to-see oval-shaped crater in the southern hemisphere called Humboldt and a mare called Humboldtianum on the north side. This crater is named for the famous eighteenth-century naturalist and geographer who, among many other achievements, studied the relationship between mountain height and falling temperature. Similar to many features that are close to the Moon's edge (or limb, as we call it), the Humboldts appear elongated. If we were to fly directly above them, they would appear more circular.

Day 2:   The beautiful walls of Mare Crisium, the Sea of Crises, are entering the morning sunlight on this day, although the floor might still be in partial darkness. Also, this day gives us our first view of the big crater Petavius, named for the seventeenth-century French Jesuit theologian. Petavius has a beautiful central complex of mountains, which towers more than 8,000 feet above the crater floor.

Day 3:   On this night, the crescent Moon is wide enough that we can see all of Mare Crisium—walls, floor, everything. We even get our first view of Picard, a crater standing in Crisium's floor. This crater proudly honors Picard, the man who founded the Paris Observatory in the late seventeenth century. Now that the Moon is high in the western sky after dark, we can see earthshine at its best.

Day 4:   Tonight the regions all around Mare Crisium are bathed in sunlight. We can imagine climbing aboard an imaginary lunar rover and taking a tour. First let's revisit Picard; toward the northern end of the crater is an "island," a crater named Peirce, which was formed long after the Mare Crisium's lava hardened. On the opposite side of Mare Crisium from the Moon's limb is Proclus, one of the most radiant craters on the Moon. This 20-mile-wide feature is named after the fifth-century neo-Platonic philosopher. Although it is small, Proclus has a system of bright rays that make it easy to spot.

On this day we can see a very old feature, the "ghost" crater Janssen, named for the nineteenth-century French astronomer who discovered the presence of helium in the Sun—the first-ever detection of helium anywhere. Some 100 miles across, the crater was formed so long ago that its story has been "written over" by more recent impact events. Its walls have been torn asunder, and it no longer looks like the fully formed crater it must have been billions of years ago. Proclus, with its system of rays still intact, is a much younger crater.

Mare Tranquilitatis, the Sea of Tranquility, is just starting to become sunlit, as is Mare Nectaris, the Sea of Nectar.

DAY 5:   This is the night for Theophilus, one of the prettiest craters on the Moon. Named for the Greek to whom the Gospel according to St. Luke and the Acts of the Apostles were written, this crater straddles the border between the Sea of Tranquility and the Sea of Nectar, two maria in full view tonight. Theophilus has a beautiful collection of peaks in its center, and it is bordered by high mountain walls. South of it is a strange feature called the Atlai Scarp, which in the Moon's infancy was a fault line. At its southeast end is a sharply defined crater called Piccolomini. This crater is probably named for the well-known Italian family with two especially famous members, one who became a distinguished seventeenth-century general in Spain and the other who became Pope Pius II.

DAY 6:   Starting tonight and extending through full phase, some maria and craters combine to form an image of a *Woman in the Moon*, whose body is centered in the Sea of Tranquility. The Mare Serenitatis, the Sea of Serenity, is her head; her legs are Mare Nectaris and Mare Foecunditatis (Fertility); Mare Crisium represents the purse she carries; and the crater Proclus, which you need binoculars to see, is her diamond ring. Once you recognize this feature you will find it an easy way to get around a sizable portion of the Moon.

Now the Mare Serenitatis reveals itself to the rays of the rising Sun. Meanwhile the Sea of Tranquility is fully visible. It was in this sea that, on July 20, 1969, humans first set foot on another world, as Apollo 11 triumphantly landed. For everyone living at the time, it was a night never to be forgotten. David Levy was in an auditorium with more than 100 children at Camp Minnowbrook in Lake Placid, New York, all watching a tiny 12-inch black-and-white television set mounted on the stage. For several hours, even the youngest children sat transfixed as Neil Armstrong stepped down the ladder onto the dusty surface of a world so ancient and yet new to us.

DAY 7:   Tonight the ancient crater Hipparchus is in full view. This is the only night on which it can be seen easily. Earlier it was shrouded in darkness, and as the Sun climbs it will cease to be prominent. But when the Sun shines at an acute angle, the crater's subtle features stand out in sharp relief. Hipparchus is named after the earliest known astronomer who kept systematic observing records, a tradition we are trying to instill in young observers even today. He also determined the true length of Earth's year and discovered the slow wobble that the Earth undergoes over thousands of years. We call this wobble *precession*, and its main effect is that as the sky slowly shifts, the pole star changes. When the Egyptians built their pyramids to point at the pole star, the star at the pole was not Polaris but Thuban, in Draco.

Inside the walls of the old crater Hipparchus are two newer craters, one of which is named Halley after the famous mathematician who predicted the return, in 1758, of the comet that now bears his name. Also tonight, two small seas have appeared, the Mare Vaporum and an area right in the middle of the Moon called Sinus Medii (Central Bay), a small flat region, which has been called the Moon's belly button.

DAY 8:   One night after first quarter, day 8 is one of the most dramatic times of the month for observing the Moon. The Apennine mountain range—the Moon's most splendid—is in full view, and the great Mare Imbrium—the Sea of Rains—is becoming more visible each night. Archimedes, a beautiful crater, is visible now, as are two smaller craters with which it forms a triangle, the large Aristillus and the smaller named for Autolycus. Born in 300 B.C., Autolycus wrote of the revolution of the spheres. Near the Moon's northern edge, two lovely features are now in view—the large lava-filled crater Plato (such features are called walled plains instead of craters) and the brightly shining peak of Pico, one of the Moon's highest mountains.

This is also the first night for the large crater Alphonsus, site of the Ranger 9 crash landing in March 1965. Nearby is a long escarpment or steep hill called the Straight Wall and the crater Ptolemy. Around 140 A.D., the Greek scholar Ptolemy published his *Almagest*, a work that used centuries of Babylonian observations of the motions of the planets to buttress his argument that the universe was centered about the Earth. As did many astronomers of his time, Ptolemy also believed that the motions of the planets could influence human behavior. He was astute enough, however, to separate his predictions of planet motions from those of human destiny.

DAY 9:   Two of the Moon's most famous craters, Copernicus and Tycho, come dramatically into view tonight. Copernicus honors the memory of the Renaissance mathematician who, on his death-bed in 1543, published his new view of the planetary system: no longer was the Earth the center of things; rather, the planets revolve about the Sun. In fact, the word "revolution," as it is applied to politics, got its start from the scientific overhaul begun by Copernicus.

The other crater, Tycho, is named after the famous Renaissance observer who discovered the supernova—the exploding star—of 1572, the first evidence in hundreds of years that the stars do change. One of the Moon's youngest craters, Tycho was carved out of a blast of a comet or asteroid that hit the Moon some 100 million years ago. This impact took place about 35 million years earlier than the one that probably killed the dinosaurs on Earth. That seems like a long interval, but in the several-billion-year history of the Earth and Moon, it is not.

South of Tycho is a large walled plain called Clavius. Named for the Bavarian Jesuit astronomer whose work led to our Gregorian calendar, this is the crater on which a lunar base supposedly existed in the movie *2001: A Space Odyssey*.

DAY 10:   The Moon is now in a gibbous phase, so prominent that the rest of the sky is also getting noticeably brighter. Some large mountain ranges and ridges, such as Straight Range and Jura Mountains, are becoming prominent. Also, Tycho's system of rays is more visible. In chapter 10 there is an experiment that should reproduce the means by which rocks and stones are thrust out of an impact basin to form a system of rays.

DAY 11:   Tonight, it is getting harder to distinguish features as the Moon continues to brighten. The rays of Tycho and Copernicus are visible even through binoculars. The Oceanus Procellarum, Ocean of Storms, is announcing its presence as the largest of the lunar plains on the side facing the Earth. Some exciting craters are coming into view as well. Kepler, a bright crater with a system of rays, is easy to see despite its small 20-mile diameter. Johannes Kepler was the scientist who discovered the supernova of 1604 and a bright comet in 1618. His main contribution, however, was his determination that the orbits of the planets are elliptical.

North and west of Kepler is the brightest spot on the Moon, the crater Aristarchus. Named for Aristarchus of Samos, who first estimated the relative sizes of Earth, Moon, and Sun, this crater is a fabulous sight in a telescope. Just north of it lies Schroeter's Valley, a spectacular feature whose southern end looks a little like the head of a cobra.

DAY 12:   On this night, just two days before full phase, the Moon is just a little more than half as bright as it will be when full. As the Moon reaches full, all the shadows disappear and its features become more difficult to see. Because there are no shadows, the Moon is much brighter when it is full than even a few days before or after full. Also, because the Moon is roughly opposite the Sun in the sky around this time, scientists call this the opposition effect. The most prominent feature tonight is the "upside down" crater called Wargentin. This crater's story is great for children; here is the way we tell it:

> The Moon has over 300 craters that can be seen through a small telescope. Some are small, others are large. Most are round, although some are oval. Some even have mountains inside, while others do not. But they all have one thing in common. They are all right side up! Except for Wargentin.
> Named after the Swedish astronomer Pehr Wargentin, who lived more than two centuries ago, this crater probably started out like most of the others. Billions of years ago a huge asteroid or comet slammed into the Moon, forming a beautiful crater surrounded by high walls. But shortly after this happened, melted rock, or lava, started to rise up inside the crater. Higher and higher the lava rose, until it almost spilled out over the crater's rim. If that had happened all we would see today is a large mound. And that would not be very odd.

But instead the lava stopped flowing just as it got to the crater's rim. Gradually it hardened as the crater walls held the lava in. So now, just a couple of days before full Moon, we can see a large group of normal craters—and one, flat as a pancake, that looks as though it is upside down.

DAY 13: As the Moon approaches full phase, the rays of the younger craters become easier to spot. Grimaldi, a small mare—a lava plain—comes into view tonight. Francesco Maria Grimaldi, a seventeenth-century physicist from Bologna, made a chart of the Moon, and we owe many of the earliest names of lunar features to him. He also discovered the dispersion of light as it passes through a prism.

There is also Pythagoras, a large crater so close to the Moon's edge that it appears to be oval. The crater was named to honor the Greek mathematician who, in addition to proposing his famous theorem, also first suggested that the Earth is shaped like a sphere.

DAY 14: Full Moon. The Moon is so bright that only the brighter stars are visible in city or country sky. Even the Moon's own features are not clearly visible, and the best things to look for are the ray systems of the younger craters.

If you want to see the traditional wide-eyed Man in the Moon, this is the best night to try. Mare Imbrium is one eye, and the other is a combination of the Mare Serenitatis and Mare Tranquilitatis. The nose is comprised of the Sinus Medii and Sinus Aestuum (Seething Bay) and Mare Nubium (Sea of Clouds) is his mouth.

DAY 15: On this interesting day, sunset begins on the Moon's east side. Features such as Mare Crisium and the crater Theophilus begin to become more prominent once again as the Sun's rays hitting at a low angle there light them up from the other direction.

DAY 16: The mountains that circle Mare Crisium are beginning to cast shadows as the setting Sun accentuates them on this night. With each passing night, the Moon will rise later and later, so fewer children—and fewer teachers—will get to see the panorama of its changing landscape. However, since it rises later, it also sets after sunrise. Thus early risers can see these features under a predawn sky.

DAY 17:   The crater Petavius is beginning to show a long shadow as the Sun sets over its high central mountain peak. Also, north of Mare Crisium are the shadowed features of some interesting craters such as Cleomedes, named for the second-century Greek astronomer who wrote about the Moon's revolutions about the Earth.

DAY 18:   Mare Crisium has all but disappeared as night begins to take over the side of the Moon that faces the Earth. A sharply visible crater named Taruntius is prominent tonight.

DAY 19:   The colossal Sea of Tranquility is shining in full relief on this night, and to the south the Pyrenees Mountains are prominent.

DAY 20:   The large crater Theophilus, with its big shadowed central peak, is a prominent feature. The Sun is also setting on a very small crater called Sabine. Just west of Sabine is the spot where humans first landed on the Moon in July 1969.

DAY 21:   The huge mountain ranges surrounding Mare Imbrium— the Apennines and the Caucasus—are easily seen on this night. Just west of the crater Calippus, named for a Greek mathematical astronomer who tried to explain the motions of the planets around 340 B.C., are the Caucasus Mountains.

DAY 22:   Tonight the Sun sets over the Apennine Mountains, and we get a good view of the mountain Piton.

DAY 23:   On this night Copernicus, Tycho, and Mare Imbrium are in full relief as the Sun sets over them. This is also a good night to view Mare Nubium.

DAY 24:   As the Moon's lit portion shrinks, earthshine is becoming prominent again. Sunset occurs over Oceanus Procellarum.

DAY 25:   The starkly beautiful Aristarchus and Cobra Head form a striking feature on this night. By now, of course, the Moon does not rise until it is nearly morning, so one must get up very early to see it.

DAY 26:   The Moon will rise in morning twilight as a very thin crescent. Small Mare Grimaldi is an easy target for observation this morning. It appears elongated because it is near the visible edge of the Moon; thus it appears foreshortened.

DAY 27:   Besides a long, oval-shaped walled plain called Otto Struve, there is not much left to see on the Moon this night. The crater is named after the nineteenth-century double-star astronomer, and perhaps also his grandson of the same name, a twentieth-century specialist in the evolution of the stars.

DAY 28:   In August and September, the waning Moon might still rise far enough from the Sun that an astute observer might get to see it. But for the rest of us, the long month's journey is over. We've taken a nice tour of our neighbor world.

Observing the Moon for a "Moonth"

(*An activity designed by Stephen Edberg, a well-known planetary scientist at Caltech's Jet Propulsion Laboratory*)

PURPOSE: To sharpen powers of observation and use them to address observational questions.

OBJECTIVE: Minimum—a daily sketch of lunar phases.

EVALUATION: A complete set of drawings of the lunar phases.

MATERIALS: Notebook, pencil. To address the questions some additional observations or measurement approaches are necessary: stopwatch or clock, binoculars or telescope (or very sharp powers of observation).

PROCESS: Students will observe the Moon daily for a minimum of 30 consecutive days (weather permitting). Students will sketch the lunar phase, providing at least an outline of the Moon's shape. The more motivated will sketch the Moon's surface and use optical aid.

QUESTIONS: (1) Why does the shape of the Moon change? (2) What causes earthshine? (3) What is illuminating the Moon? (4) What phases can be seen during which times of day? (5) What is the geometry between Illuminator–Moon–Observer at various phases? (6) Is the Moon rotating as it goes around the Earth? (7) How long is the period of rotation?

VOCABULARY: Phase, earthshine, rotation, revolution, elliptical orbit.

CHALLENGE: By what observation techniques can the questions be answered?

OBSERVING THE MOON IN DAYTIME

Fair, but of fairness as a vision dream'd;
Dry were her sad eyes that would fain have stream'd;
She stood before a light not hers, and seem'd

The lorn Moon, pale with piteous dismay,
Who rising late had miss'd her painful way
In wandering until broad light of day;

Then was discover'd in the pathless sky,
White-faced, as one in sad assay to fly
Who asks not life but only place to die.

—Gerard Manley Hopkins, *A Fragment of Anything You Like*

Activity

The Sun is up only in the daytime, and the Moon is up only at night, right? Wrong! The Sun's presence defines the daytime sky, and its absence defines the nighttime sky. It turns out, however, that half the time the Moon spends above the horizon, it is daytime. This fact opens the possibility of observing the Moon with binoculars and a telescope during school hours. We recommend the following steps:

MATERIALS:
Telescope   Chair for shorter people
Binoculars   Observing logs

PROCEDURE:

1. Determine when the Moon is likely to be visible in daylight. The best time is around last quarter phase, when it can be seen in the west during the morning hours. When the Moon is waxing toward first quarter, it rises in the east in the afternoon, and will probably not get high enough to be observed conveniently until after school has ended for the day.
2. Explain that the Moon is not always a nighttime object and that it is often visible during daytime.
3. Explain why we are looking at the Moon. Using the thoughts and information provided earlier in this chapter, emphasize: (a) that it is another world, (b) that it is a world in our neighborhood, and (c) that it has craters that tell its story, as well as giving us strong hints about the corresponding history of our own world.
4. Set up a small telescope and as many pairs of binoculars as you can get. Have the children look through both telescope and binoculars and compare the views. The binocular view will be different from the view through the telescope. Because the binoculars show so much more sky around the Moon, there is more contrast between the Moon and the sky: The Moon shows up more clearly through binoculars than it does through a telescope. However, the telescope will magnify the Moon more and it will invert it, making it appear upside down.
5. Make sure each child gets a good, comfortable view. Don't rush this look. Ask each child to describe the sight.

CLOSURE: Discuss the meaning of what the children saw: a distant place that 12 people have walked on and have taken rocks from. Ask them to write up their observations in their observing logs.

WHEN THE MOON IS ECLIPSED

During the course of the Moon's travels around the Earth, there is only one phase—full Moon—when an eclipse of the Moon can occur. When the Earth is directly between the Sun and Moon, its shadow can fall on the Moon, causing the eclipse.

So much has been said about the dangers of observing the Sun when it is undergoing an eclipse that some people fear that a lunar eclipse is dangerous to view, too. Not so: a *lunar* eclipse is entirely safe to watch with the unaided eye, binoculars, or a telescope.

As the start time of a lunar eclipse approaches, you might notice that the full Moon's brilliance has started to diminish. You're right: the Earth has two shadows, an inner shadow or umbra and an outer shadow called a penumbra. The umbra is the Earth's central shadow, in which the Earth completely covers the Sun. The penumbra is the shadow that the Earth casts when it only partially covers the Sun. As the penumbra advances before the start of the main eclipse, the Moon gets a little darker. The "first contact" with the Earth's inner shadow takes a sharp, very noticeable bite, and as the minutes pass the Moon changes its appearance dramatically as the bite goes deeper. If the eclipse is total, the shadow will march all the way across the Moon during the next hour, changing the appearance of the Moon from its usual dazzling white to a much darker color. Totality could last from about 30 to more than 90 minutes. Then, as the eastward-moving Moon moves away from the shadow of the Earth, the features return one by one to sunlight.

The biggest uncertainty in any lunar eclipse, besides the clouds, which might not allow us to see it at all, is how dark the eclipse will be. Total lunar eclipses can darken the Moon from a bright orange to a very dark brown. How dark an eclipse is depends on the amount of dust in the Earth's own atmosphere. The lunar eclipse on December 30, 1963, closely followed the eruption of a volcano that threw a great quantity of dust into the Earth's upper atmosphere, making the Earth's shadow especially

Lunar Eclipses

Date	Type	Approximate area covered
September 16, 1997	Total	Australasia, Asia, Africa, Europe
March 13, 1998	Penumbral	North and South America, Africa
August 8, 1998	Penumbral	North and South America, Africa
September 6, 1998	Penumbral	Asia, Australia, North and South America
January 31, 1999	Penumbral	Africa, Asia, western North America
July 28, 1999	Partial	Asia, Africa, North and South America
January 21, 2000	Total	North and South America, Europe, Africa
July 16, 2000	Total	Asia, Australia, western North America

dark. During the total portion the Earth's shadow was so dark that the Moon was almost invisible to the naked eye and barely visible even through a pair of binoculars.

For the benefit of children getting their first look at the sky, we are very lucky to have the Moon. No other planet besides Pluto has such a bright moon in its sky. The Moon enables children to visit the features of another world—its plains, craters, mountains, and valleys.

There are few things more exciting in astronomy than a look at the Moon through a telescope. Its craters are named for people who represent milestones in the history of astronomy and its related sciences. People have admired the Moon for thousands of years, but now we see a Moon that people have landed on. The Moon landers have collected its rocks and brought them back to Earth, where they are now displayed in museums and planetariums in many cities. The idea that this world that we can see tonight from our backyards is a world that people have walked on and driven across is thrilling.

The Moon is a personal place. As children look at it for the first time through a telescope, they should let their imaginations

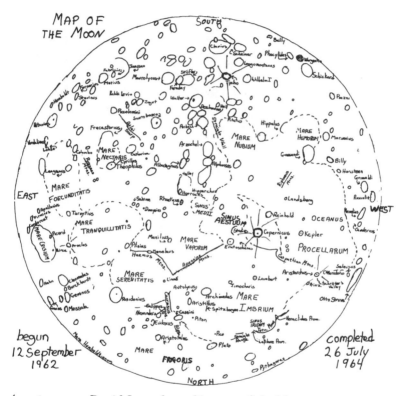

As a teenager, David Levy drew this map of the Moon over a two year period from observations he made with some small telescopes. The map is adapted from an earlier version printed in his book The Sky: A User's Guide and is used by permission of Cambridge University Press.

roam across the heavens and think what it would be like if they were really there. Because the Moon's gravity is so much less than the Earth's, a child's Moon weight would be only a sixth of the Earth weight; a child weighing 70 pounds would weigh in at only 12 pounds on the Moon. That means that a walk of exploration could turn into a romp up steep hills and down crater walls, encumbered only by the extra weight of the space suit. That suit would be all-important. It needs to protect its wearer against temperatures that at midday would top 200 Fahrenheit degrees above zero and at night would plunge that far below zero. It is a very hostile world to walk on. But it is possible to visit this world; people have done it. In their imaginations, children can too, right from the eyepiece of a telescope.

Chapter 9

Planets as Places

If a long-lost cousin from Venus were to visit, how warm would we have to keep the house? If we tuned into radio station MARS, what would the weather report be like? On Jupiter, would our enormous weight allow us to do pushups? How long would we have to wait in cold darkness before we celebrated our first birthday on Pluto?

The answers to these simple questions give us a stunning early revelation of what our neighbor worlds in the solar system would be like. First, the visitor from Venus: He would actually freeze to death in a second in our house, even if the heat were turned on as high as it would go. He would even freeze to death in a few minutes if we put him inside *an oven set to broiling*. To make him comfortable, we would have to double the temperature in the oven, to about 800 degrees Fahrenheit!

Now let's tune into station MARS. We hear the forecast, and at first it looks good: "Pleasant sunny day, beautiful pinkish sky. High about 65 degrees Fahrenheit." That seems pleasant enough, but we listen some more. "Clear and cold tonight, with temperatures falling to 50 below 0." And that's not all. The beautiful sunny temperatures, we learn, only hold if we are an inch or two from the ground, and near the equator. The Martian atmosphere is so thin that with every foot we climb, the temperature falls drastically.

Moving on to Jupiter, a 100-pound person would weigh 250 pounds while standing on that planet. But that person couldn't stand anywhere, for Jupiter is virtually a large gas bag. If there were a solid surface, the child would weigh so much that walking—even crawling—might be impossible. The gravity of the planet would exert so much pressure on a person that he or she could barely breathe.

Finally comes Pluto, the little planet with the big moon that takes 248 years to orbit the Sun. Someone born on Pluto would live out a life and pass away long before the planet made half an orbit about the Sun. There would never be a birthday party on Pluto.

Such is the variety of conditions on the worlds in our solar system. As a teacher or a parent, you have the chance to present these exciting ideas about this wealth of different places. Although we have never sent people to these planets, spacecraft have visited them. The intrepid spacecraft Voyager 2, in its odyssey that began in 1977, took pictures and recorded information as it passed Jupiter, Saturn, Uranus, and Neptune. Like the Moon, these planets are places we have seen and where children can dream of visiting.

PLANETS OF ROCK AND GAS

Although the nine planets offer a variety of environments, they are generally of two types: the inner ones with rocky surfaces and the outer gas giants. These are very different kinds of planets. The inner ones are much smaller than the outer ones, and they all have atmospheres—those of Mercury and Mars being quite thin and those of Venus and Earth much denser.

MERCURY

At first glance, this picture looks like the Moon taken through a large telescope. But it is a photograph of Mercury taken in 1974 by the spacecraft Mariner 10 as it made two passes by the closest planet to the Sun (see Figure 37). Mercury has many craters on its surface, the result of impacts by comets or asteroids over billions of years. However, some areas have many small craters, implying that at one time lava flowed over other parts of Mercury's surface. Also, Mercury has many large cliffs, which are thought to have formed as the interior of the planet cooled and shrank inside an already solid crust, like a dried-out apple with a crinkled skin.

VENUS

Exploring Venus

In August 1962, a tiny spacecraft called Mariner 2 rode atop an Atlas rocket from its launch pad at Cape Canaveral, Florida (now called the Kennedy Space Center). The ride through the atmosphere was rougher than expected. As the rocket boosted its precious cargo through the atmosphere, it started to turn slowly on its axis. The launch crew was worried. If the rolling continued, the booster could suddenly lose track of where it was going and wander off course. There were so many launch failures at the time that another one wouldn't be surprising. After all, Mariner 1 had failed just a few months earlier, its load of expensive scientific instruments falling uselessly into the Atlantic Ocean.

Then the rolling stopped. With all these accidental failures, one NASA scientist quipped, now we have an accidental success! The booster completed its work, and the spacecraft went on to succeed as the first mission to another planet. In December 1962, Mariner 2 sent back signals from Venus. We learned that day that Venus's temperatures, near 800 degrees Fahrenheit, make it a forbidding place. Mariner 2 was able to study Venus's cloud tops, but it could not see the planet's surface.

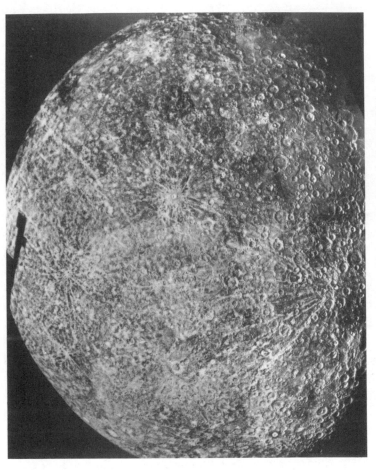

37. *Mercury as seen from Mariner 10, March 1974. NASA image.*

More than 20 years later, the U.S. spacecraft named after the explorer Magellan followed the course of Mariner 2, but it paid Venus a much longer visit. Instead of the quick flyby of earlier spacecraft (and some spacecraft from the former Soviet Union even landed, frying away in the heat within a short time) Magellan settled into an orbit around Venus. The craft was equipped with a radar mapping device. Instead of minutes, Magellan had years to complete its mapping.

Observing Venus

Although Venus is the brightest "star" in the sky, it is never far from the Sun. Either visible in the evening sky after sunset or in the morning sky before dawn, Venus is a strangely compelling object. In ancient times, Greek observers called it Phosphor in the evening sky and Hesper when it was in the morning sky. The famous nineteenth-century American writer Henry David Thoreau wrote of the "Hesperian Isles," in a sense the islands of Venus, the guardian of night: "Far in this ethereal sea lie the Hesperian isles, unseen by day, but when the darkness comes their fires are seen from this shore, as Columbus saw the fires of San Salvador." We think that these islands of Hesper beckon all of us. According to one interpretation, they were the daughters of Venus, who lived in a garden at the extreme western end of the world and guarded the celestial flock of night.

Whatever the interpretation, Venus never gets far from the Sun in the sky because it is physically closer to the Sun than Earth is. And although Venus orbits the Sun in about two-thirds of an Earth year, it actually takes more than eighteen months for Venus to go once around the Sun relative to the Earth, which of course is also moving around the Sun. Consequently, Venus spends several months in the evening sky, followed by several weeks too close to the Sun to be seen, followed by several months in the morning sky.

Through a telescope, Venus appears as a blank white disk which, like the Moon, shows a phase. Venus is brightest when it is a thin crescent. If Venus is in the evening sky during an appropriate season, you can observe it every week or two and watch it slowly change its phase.

Activity: Watching Venus's Phases

PURPOSE: To observe the changing phases of Venus.

MATERIALS:
  Telescope
  Observing log

PROCEDURE: If Venus is visible in the evening sky, use a telescope with a low-power eyepiece to observe our sister planet. Make one observation each week if weather permits. You can infer the positions of the major planets Venus, Mars, Jupiter, and Saturn from tables supplied in major astronomical magazines such as *Astronomy* and *Sky and Telescope*. Have the children draw the phase in their observing logs.

CLOSURE: Ask the children some questions: What are the names of the nine planets, which ones are closer to the Sun than Earth, and which ones are farther away?

Which planets show phases? The answer is that, generally, only the planets closer to the Sun than Earth show the effects of phase. In 1994, the Galileo spacecraft photographed Comet Shoemaker–Levy 9 as it crashed into Jupiter. The resulting pictures showed Jupiter in a gibbous phase, with the bright flashes of the impacts appearing on the dark part. When we look at Jupiter, or any remote planet, from the Earth, it always appears at close to full phase. But at the time, Galileo, though still far from Jupiter, was almost at Jupiter's distance from us. From the spacecraft's point of view, Jupiter appeared in a gibbous phase. Whether a planet shows a phase depends on where we are observing it from. Were we on Jupiter, all the inner planets, including Earth and Mars, would appear to go through phases as they orbited the Sun.

Transits of Venus

At rare intervals, Mercury and Venus pass directly between the Sun and Earth. Such an event is called a transit. With Mercury, transits occur occasionally when, for an hour or so, the tiny planet appears as a dark dot crossing the face of the Sun. Transits of Venus occur only twice a century, in pairs 8 years apart. It is the differing tilts of the orbits of Mercury, Venus, and Earth that make these transits so infrequent.

When the next transit of Venus occurs in the year 2004, people will travel great distances to see it. Transits of Venus are a part of our historical legacy, as they have helped us determine Earth's distance from the Sun. Those who miss this transit will have a chance to see the following one, which will take place 8 years later in 2012. But after that, there won't be another one until well into the twenty-second century.

All these events—phases, transits, appearances of planets in the morning or evening sky—help to show that the solar system is working like a giant clock with the Sun in its center and all the planets and moons orbiting around it. To see any of these things actually happening is a marvelous experience that makes us feel as though we are actually a part of the action.

Activity: The Venus Topography Box

BACKGROUND: The Venus Topography Box is an activity similar to one developed in the 1950s. A similar activity, Sensing the Unseen, Mapping the Ocean Floor, was part of the NSF material for National Science and Technology Week 1992–1993. It is an activity that can be used to show how scientists can map any hidden surface, such as the cloud-covered surface of Venus (using radar) on the ocean floor (using sonar). An imaging radar similar to the Magellan radar was flown on the Shuttle in 1994 to map Earth's surface. Other missions will use laser altimeters along with images to map the surface of Mars.

When we go into a classroom, we provide premade boxes.

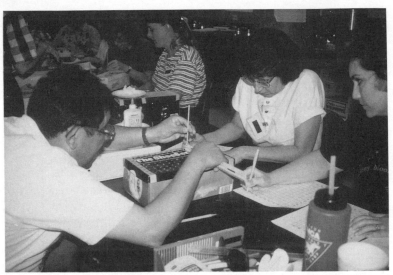

38. *Venus Topography Box. Phred Salazar, Anita Mendoza, and Rosemarie Hagstrom.*

Many teachers cannot manage the cost, time, and mess of plaster-of-Paris, which would enable students to make their own boxes. We can tailor the number of holes and the size of the graph paper to the grade/skill level of the students (see the discussion below). Following is a description of how we make our presentation in a classroom:

1. The class is divided into groups of about five to six students. When we go into a classroom situation where we have only an hour or so, we find that two of us plus the teacher can work well with up to about six groups. However, we know of several teachers who, given enough time and student preparation, have done this activity on their own with their classes.
2. Ask each group to provide a ruler; lead pencils; colored crayons, pencils, or markers (colors from color chart).
3. Provide each group with a closed topography box, measuring stick, and graph paper.

CONSTRUCTION:

1. Crumple newspaper, lay in shoebox to form an uneven "surface" (mountains, plains, valleys).
2. Cover newspaper with aluminum foil.
3. Pour plaster of Paris over the foil.
4. Use a pen to draw a grid on the shoebox lid.
5. Use a large nail (or ice pick or pencil) to punch holes in the lid.
6. Letter and number the grid.

ALTERNATIVE FOR PLASTIC BOXES:

1. Line the bottom of the box with a sheet of styrofoam, use other pieces to make a crude "surface."
2. Pour white stucco, plaster, or cement over the styrofoam.
3. The stucco can be "molded" to form craters, mountains (volcanoes), valleys, etc.
4. Draw a grid on the box lid or tape a sheet of graph paper on the lid (do not do this with the hot awl).
5. Use a heated awl or drill to make holes in the lid.
6. Letter and number grid.

Other than the cost of the plastic box (one can use a shoebox), this is cheaper and stronger than the plaster-of-Paris method. Also, since the plaster-of-Paris turns hard within 5–10 minutes (vs. 30 minutes for the stucco or cement and longer for plaster) you have more time to work the stucco into topographic features. The stucco is also easier to repair (the old plaster-of-Paris absorbs the water from the new plaster and so the new plaster hardens almost immediately).

PROCEDURE:

1. Describe radar, radar mapping, Magellan mission. For elementary school classes, you do not need to go into the details of the actual mapping techniques used by Magellan (delay-Doppler mapping and a radar altimeter). You can also describe sonar mapping of the ocean floor, sonar to detect sub-

marines and radar as used at airports as well as the radar used by bats and dolphins. You can model time vs. distance with a tennis ball, bouncing it on a desk or file cabinet vs. the floor. When we do this we bounce the ball on a table and then ask the students if it will take less, the same, or more time to bounce back when we bounce it off the floor (more time because the distance is greater).

2. We explain to the students that we cannot bring a radar into the classroom, so we will model radar and the surface of Venus.

3. We then tell the class that they represent a scientific research corporation (or university) specializing in radar mapping. NASA wants to land a spacecraft on an unseen (because of clouds) section of Venus. It is their job to determine the best area to land. Ask what kind of area would be best to land a spacecraft (flat).

4. Describe how to insert the measuring stick and record data. Do one measurement together. If there is enough time to have the students do their own measurements with a ruler, discuss rounding up (to an inch, a centimeter, or a half centimeter as appropriate). Begin measurements. Note that for very young students, you can use a color-coded stick. A stick that is pre-marked and numbered can be used if the teacher feels that the children might have trouble measuring with a ruler. When we go into classrooms, we use the last method since it saves a lot of time. Our increments are 1 centimeter, which will give about 10 divisions for a typical shoebox.

5. Adult helpers should be sure tasks are being shared and data are in the correct squares. As the activity proceeds, ask the students in each group if a small number is a high spot or a low spot. Ask them to predict what the next number will be. As patterns develop, ask them what they are seeing (valleys, mountains, volcanoes, etc).

6. At this point, older students will usually ask "What is the scale? How big is the spacecraft?" If your group does not, set a size such as, "We need a square of at least nine boxes for a safe

landing spot. Be sure your flat area is big enough." In one class, a student wanted to know if we could use a plane. On Venus, it might be better to use a planelike lander, requiring a long landing area.

7. Once all measurements have been recorded, show the group(s) how to organize their height measurements into a topographic map. This will involve an outlining of squares in most grades. If time permits, older students can be shown a real topographic map and you can explain contour lines. They can then try to do this on their Venus graphs. In this case have them use the graph paper with the larger number of squares so that they can number every other box and draw the contour line between the numbered boxes.

8. Distribute the color key and allow students to color in their maps. Each group should determine the best (if any) area on their map for landing. A representative from each group should present their case to the class. A vote can be taken to determine the best landing site. (They should now open their boxes to check for accuracy.)

9. Brainstorm on how to get finer resolution measurements (it will surprise the students to see how rough an apparently flat area can be). Finer measurements can be made on the stick for greater altitude resolution or more holes can be drilled in the box lid (twice as many holes across and down, however, means four times as many holes).

10. A 3-D model can be constructed out of clay by converting the measurements into a model that can then be compared to the actual surface. Sugar cubes also can be used, remembering that a 1 means a high spot so you cannot just stack the same number of cubes as the number in a given grid square.

11. You can then discuss the real-life situation of how scientists and engineers determine where to land on a planet: the scientists want to land at an exciting site (e.g., in ancient river valley or near a volcano), whereas the engineers want to land in a smooth, flat place since they have the safety of the spacecraft in mind, even though it may be "boring" scientifically.

EARTH

The Earth as a Planet

Because we live here, it is very hard to visualize the Earth as a world orbiting the Sun. That is why it took our ancestors so long to understand it. In earlier times, ideas as far-fetched as the Earth being supported by four elephants standing on the back of a turtle were called up to explain our place in the scheme of things. Consequences of that theory, for example, just where the turtle stood, were not gone into. Not until 1543 did Nicolas Copernicus publish, on the last day of his life, his treatise *De Revolutionibus* about how the Earth and the other planets orbit the Sun in space. More than 60 years later, Galileo actually proved this theory through his observations of four moons orbiting Jupiter and the phases of Venus.

What evidence do we have, during our everyday activities, that the Earth is a planet orbiting the Sun at 18 miles a second? For many children, the problem with this idea is that if the Earth is really racing along so fast, it should be extremely breezy outdoors. Some children are not entirely convinced by the explanation that the Earth moves through a vacuum of space, carrying its atmosphere of air along with it.

1. Day and Night. The best evidence that the Earth is a world in space comes as the Sun appears to rise and set. The change from day to night every day is a very dramatic thing. Here on Earth, day and night are so much a part of our natural experience that we don't notice the change, or certainly don't make a fuss about it. Isaac Asimov made this point in his classic science fiction story *Nightfall*, in which a planet going around a group of suns experiences a rare line-up when all but one of the suns is set, and the remaining one is eclipsed by a moon. As the people see a brief period of night, the gradual vanishing of the light results in everyone's panicking, except, of course, the astronomers holed up in the observatory and a religious sect that has kept a record of the previous "nightfalls." The apparent movement of the Moon and stars across the sky is the same thing as sunrise and sunset, and is a further manifestation of the Earth turning on its axis.

2. *Tides—Part a.* The phenomenon of the rise and fall of ocean waters twice each day offers a different kind of evidence that the Earth is rotating.

3. *The Moon Moves from Night to Night.* Each night, the Moon appears in a different place in the sky, a certain distance east of where it was the night before. This change is readily seen in both the position of the Moon among the stars and the changing times of moonrise or moonset, and is evidence that the Moon is orbiting around the Earth.

4. *Tides—Part b.* Each day high and low tides occur considerably later. This effect is due to the Moon's changing position as it orbits the Earth.

5. *Seasons.* In a film called *A Private Universe,* several Harvard students dressed in full academic regalia and holding their newly awarded degrees were asked if they knew the cause of the seasons. Most graduates interviewed had no idea of the correct answer: They said that the Earth is closer to the Sun in summer than it is in winter. Although the Harvard graduates imagined that the Earth's orbit is an ellipse, and that the Earth is much closer to the Sun in summer than it is in winter, that perception is not true. The Earth's orbit is almost a perfect circle, although we actually are closest to the Sun on January 2—the middle of winter in the Northern Hemisphere. The Earth rotates and it orbits the Sun. The Earth's axis of rotation is tilted; because of this tilt, parts of the Earth away from the equator are tipped toward the Sun during half of each orbit and away during the other half. The seasons change because of the way the Earth is tilted as it orbits the Sun. Because the Earth's Northern Hemisphere tilts away from the Sun in January, that half experiences winter at that time. The Southern Hemisphere, which tilts toward the Sun at that time, has summer.

6. *Changing Night Sky.* Because the Earth revolves about the Sun, each night we get a slightly different view of the sky. From night to night the change is not much, as the stars rise about 4 minutes earlier. But that quickly adds up—a half hour a week, 2 hours with each passing month where, at the same hour, the sky looks different. The change happens because we look at it from a different perspective as the Earth orbits the Sun.

7. Meteors Falling. It is easy to see the bright flashes in the night—what many people incorrectly call shooting stars—that come when small particles of dust, mostly debris from comets, enter the Earth's atmosphere. In its annual journey, the Earth encounters several swarms of dust. The richest occur on August 11 and December 12, and at these times you can see these interactions of small particles with the Earth's atmosphere, if the sky is dark, as often as once every few minutes.

8. The Dance of the Planets. The planets move slowly through the sky, and it is easy to see this motion during a season of observing. This movement is due to the planets orbiting the Sun. However, as the Earth overtakes the more distant, and slower moving, planets, these appear to slow down, reverse direction for a few months, and then resume their eastward journey. The effect is similar to passing a slower car on a highway; as we overtake the other car, it appears to move backward. It is this sense of retrograde motion that gives us evidence that the Earth is also moving through space.

9. Appearance and Changing Position of Comets. As a comet moves past the Sun, it grows a coma and tail, which seem to change shape. Part of the reason for this change is that we are observing the comet from different angles as it and the Earth move about the Sun.

The Earth as a Book

The Earth does have a long story to tell, a story written in pages of rock. Layers of rock deposited over billions of years can tell us when areas were under the sea and how they eventually were lifted up to become vast mountain ranges. Although we see the very beautiful and varied surface, with mountains, plains, and oceans, that makes up the Earth's crust as we know it today, the bulk of our planet is made of much softer rock called the mantle, and there is a hot core in the middle.

In the most recent billion years of our world's history, many of these rocks have contained fossils, the remains of creatures long gone, creatures whose bones have changed, molecule by molecule, into the rock that preserved them. Thanks to this process, we learned of the small trilobites that crawled about for half a billion years and of the age of the dinosaurs that ended so abruptly 65 million years ago. In a different way, the Earth preserved for us the story of their demise. In a tiny layer not much thicker in some places than a postage stamp, but which is spread out around the world, lies the soot, dust, and iridium from the impact. Buried under Mexico's Yucatán Peninsula is the vast impact crater the comet left when it hit (see chapter 10).

MARS

On one historic Halloween night in 1938, millions of people listened to Orson Welles's *Fireside Theatre* radio program about an army of Martians invading New Jersey. Although the show was a fictitious drama, it set off widespread panic among people who really believed that (a) there was intelligent, belligerent life on Mars, and (b) the Martians were actually launching an invasion of Earth. It was so realistic that even scientists were initially taken in. Clyde Tombaugh, who had discovered Pluto only 7 years earlier, was excited about the fact that we might finally be meeting extraterrestrials. But a few minutes into the program, it was reported that only two nights earlier, astronomers from a Mount Jennings Observatory had witnessed the launch of the invaders from Mars. "Wait a minute," Tombaugh mused. "Mars is not visible in the night sky now. No one could have been observing Mars from any observatory!"

More than any other remote planet, Mars has been thought to be a home to life ever since Schiapareli, in the nineteenth century, drew a network of strange lines on Mars he called *canali* or channels. Late in the nineteenth century the American astronomer Percival Lowell expanded the notion that there were artificially built canals on that distant place: He thought that they might have been constructed by a civilization desperate to conserve a dwindling water supply. He even wrote about the bravery of Martians, who had overcome their regional differences and joined together to solve a common dilemma.

In 1964, the American spacecraft Mariner 4 visited Mars and recorded a planet covered with craters. Later visits, especially by the Viking spacecraft in 1976, confirmed that the planet has a fascinating surface, which includes a canyon called Valis Marineris that dwarfs the Grand Canyon, but as of 1996 no evidence of life had yet been found.

Life on Mars?

In the summer of 1996, NASA made a stunning announcement. On a meteorite from Mars, scientists had detected possible evidence that microscopic life might have existed on the red planet more than a billion years ago.

How did the rock get here? Some 15 million years ago a comet or asteroid smashed into the surface of Mars, hurling tons of rocks into space. Orbiting the Sun, at least one of these rocks spent eons in the dark region of space between Mars and Earth, until one day some 13,000 years ago it crashed into the Alan Hills region of Antarctica. It remained undisturbed in that frozen land until 1984, when meteorite hunters on a New Year's ride found it.

The rock had suffered an earlier impact far back in Mars's history. Between 3 and 4.5 billion years ago, a comet or asteroid gouged the rock out from its placement deep beneath the surface of Mars, and allowed water, and possibly tiny forms of life, to seep into it.

The implications of the Mars rock are immense. If life began on Mars as well as Earth, does this mean that life is common throughout our galaxy? If life did evolve on Earth and Mars together, why did Mars stop being a home for life? And if it happened to Mars, could it happen to Earth?

Moons of Mars

Mars's two moons are the centerpieces of another interesting story, which began a full century and a half before they were actually discovered. In 1726 Jonathan Swift wrote in his classic *Gulliver's Travels* of the discovery of two Martian moons by a team of scientists living on a flying island. There is no possible way that Swift could have known of the two real moons of Mars, but he was correct in predicting that if Mars did have moons, they must be small or they would have already been discovered. In 1877 Asaph Hall, observing from Washington's U.S. Naval Observatory, found two real moons, which became known as Phobos, meaning fear, and Deimos, for terror.

There is another part to this story of objects that are related to Mars. In 1990 David Levy and Henry Holt discovered an interesting asteroid that is not a moon of Mars but does follow Mars around the Sun, trailing it a sixth of the way around its orbit. It is now Asteroid 5261 and is called Eureka, after Archimedes's excited words after making a discovery—"I have found it!"

Observing Mars

It is interesting to talk about Mars in the classroom, but it is often difficult to observe it through a telescope. It is the closest major planet to Earth but does not usually position itself favorably. The best way to determine when Mars is favorably placed is to read the information in newspaper columns or in magazines such as *Sky and Telescope* and *Astronomy*. Through a small telescope Mars might look like a bright red star or a small disk.

With the unaided eye, Mars does resemble a red star—the reddish tinge is easy to distinguish. It moves relatively quickly among the stars, so its motion should be easily seen over a week, especially when the planet is opposite the Sun in the sky; that is, it rises around sunset. When Mars, or any of the other outer planets, is viewed this way, we say that it is "in opposition."

Even at its best, Mars takes up less space in a telescope eyepiece than a large crater on the Moon. So it is difficult to see many features on the planet using a telescope. However, some features, such as the famous Syrtis Major and the Solis Lacus (also called the Eye of Mars), sometimes appear as dark areas on this planet. We encourage our young observers to watch Mars when the opportunity arises and to draw what they see.

The "Face" on Mars

During the middle 1980s, people looking at pictures taken of the surface of Mars by the Viking orbiter spacecraft noticed a feature that looked like a great stone face. It is merely a hill, photographed at a low Sun angle so that it appeared to take on the shape of a face. But some people have decided to capitalize on the absurd notion that the structure was built or carved by Martians. There have been interviews with "scientists" who insist that the face is intended to show a Martian civilization trying to contact us. It is not. The hill is just that—a hill. Had the picture been taken 15 minutes later, the shadows would have been very different and would not have resulted in that appearance. Mountainous features on Earth can suggest similar faces, such as the profile of Cochise in southern Arizona or New Hampshire's Great Stone Face.

JUPITER

Well named by the Romans for the king of their deities, Jupiter is the greatest of the planets. Whenever it is in the sky it shines as one of the brightest objects. It is a planet whose observation has a dramatic history, and for children it provides an excellent connection to history and sociology. We are referring here to the story of how Galileo, the famous Italian astronomer, discovered the four bright Jovian moons. Here is what happened:

In 1609 an Italian astronomer named Galileo Galilei heard of an astonishing invention from Holland. The spectacle maker Hans Lippershey had invented a wondrous device that consisted of two glass lenses placed a certain distance apart, and by looking through the two lenses, he could magnify distant objects. Galileo immediately saw the potential of this invention. He increased its power to 10, turned it toward the sky, and observed the brightest objects—the Sun, Moon, Venus, and Jupiter. Discoveries of craters on the Moon and spots on the Sun came quickly. He also discovered the phases of Venus, which proved that Venus orbited the Sun, not the Earth. Then he found a set of three stars near Jupiter. The following night Galileo saw the three stars in different positions, and a fourth appeared. Galileo concluded that these moving stars were moons that revolved around Jupiter. For the first time, worlds had been found that clearly did not revolve around the Earth. The Jovian moons were Galileo's greatest find, and one of the seminal discoveries in the history of science.

At that time, astronomers and artists alike were debating whether the Earth was the center of the solar system, with the planets and stars revolving about it, or whether the Sun was at the center. To anyone who looked through Galileo's telescope, it was obvious that the Earth-centered system could not work. Here were four bright moons that clearly revolved around the planet Jupiter.

In 1611, just 1 year after Galileo's discovery of the moons, the Catholic Church filed the first of several documents against him—in secret and apparently without his knowledge. In 1616 a document known as the *Codex* forbade any teaching that the Sun is in the center of heaven. Galileo decided to wait until a more propitious time to defend the new idea. That time came, Galileo thought, when a new Pope was elected in 1623. Urban VIII seemed open to new concepts, and in the new Pope's first years Galileo walked and talked with him. But Galileo seriously misread these discussions. In 1632 he published his *Dialogue on the Great World Systems*, in which he presented both the Sun-centered and the Earth-centered universes as a debate. Unfortunately, the character in his book who represented the Earth-centered system was named Simplicio, and he seemed to mock the Pope.

Urban VIII was outraged. He promptly put the old and virtually blind astronomer at the mercy of the Holy Office of the Inquisition. Galileo was shown the instruments of torture and during his trial in 1632 was forced to recant his observations and conclusions about the Sun being the center of the universe.

Telling this story to children is not easy, but we think that it is important. We dramatize the tale for children, but first we explain to them that it is a drama. Children are familiar with television shows that might present the truth in different ways.

That it was the Church that so controlled public thought is not as important to us as the fact that some central authority did. So instead of mentioning the Church by name, we discuss first the historic nature of Galileo's discovery, and then introduce four men in long, black robes, ceremoniously debating with Galileo their belief that everything revolves about the Earth. First, they claimed that the Moon revolves about the Earth. "Is that true?" we ask the children? Then, they insisted that the Sun revolves about the Earth. "Is that true?"

Galileo was hauled into Rome to face his trial. The children imagine a long table, at one end of which were four large thrones, one for each of the men in the long, black robes. At the table's other end was a tiny wooden chair with a leg that dangled. At the center stood Galileo's telescope.

"On that chair sat the old astronomer. 'Galileo,' chimed the men in the long, black robes, 'we offer you a choice. You can claim that your telescope is broken. You can tell us that there are no moons of Jupiter, and then you can go home. Or you can say that your telescope is fixed. And now that you have seen the instruments we use to torture people, you can imagine the consequences of saying that your telescope is fixed.'

"For a long time Galileo sat on his old wooden chair with the leg that dangled, staring first at the men in the long, black robes, then at his telescope. Then he rose. 'Gentlemen in the long, black robes,' he intoned, 'my telescope is …'"

At this point we just stop the story and wait. The tension is always high in the room as the children wonder what will become of the old astronomer. We walk toward the telescope, and continue … "my telescope is … broken! There were never four moons of Jupiter!"

CLOSURE: The discussion that follows this story is always interesting, and should be divided into both cultural and scientific parts. Cultural questions that can be raised include: Who really won this battle? Can this type of episode, where someone is harassed, imprisoned, tortured, or even killed for writing what he or she believes in, happen today? Current issues should be raised here, for example the case of Salman Rushdie, forced to live in hiding for several years after the Ayatollah Khomeini called for his death because of his novel *The Satanic Verses*.

The scientific part of the discussion is more straightforward. Who really won this battle? Centuries later, we are indeed allowed to look through telescopes at Jupiter's moons and at anything else. In fact, a spacecraft launched by the United States, but with experiments on board from other countries, is intended to study these moons. This spacecraft is named to honor Galileo. Imagine how excited he would be if he could return and see that!

An interesting point to be made to older children is that for all the fuss about whether the Earth or the Sun lies at the center of the Universe, the correct answer is neither. As the centuries passed, our understanding of the Universe put our solar system farther and farther from the center of the universe. During the second decade of this century, Harlow Shapley showed that the Sun is far from the center of the galaxy in which we live. Some 10 years later, Edwin Hubble expanded that notion to show that we live in one small area of a universe that is expanding rapidly in all directions. So the facts of the argument that raged so many centuries ago were wrong on both sides. But its meaning has lost none of its strength over all that time. Copernicus and Galileo were the first to begin the task of moving us away from an Earth-centered universe. By finding new worlds, Galileo's contribution was to help bring our world out of the Dark Ages.

SATURN

Galileo might have discovered the truth about Jupiter and its moons, but he had no idea of how to interpret what he saw when he first turned his telescope to Saturn. He thought he saw two large, misshapen moons, one on each side of the planet. They did not appear to change their position relative to Saturn. As years passed, the mystery deepened—the "moons" slowly faded and disappeared, only to reappear a year later and slowly brighten.

It was not until 1656 that Christiaan Huygens solved the riddle. A brilliant man who had invented the pendulum clock, Huygens saw that Saturn had a ring.

Even small telescopes show Saturn's exquisite rings. However, the rings have different appearances, depending on where Saturn is in its orbit about the Sun. Twice in each 29-year orbit, the rings actually disappear, as they did in 1966, 1980, and 1995. When the rings are seen edge-on, they are so thin that they seem to vanish.

Looking at Saturn and its rings offers a true highlight in astronomical observing. Through a small telescope, they appear as a single broad ring. With a larger telescope, however, you might see the faint gap in the rings. Known as Cassini's division, it is the largest breach in the rings. The Voyager 1 and 2 spacecraft that flew by Saturn have shown us that the major rings are subdivided into many hundreds or thousands of tiny ringlets. However, these divisions are not discernible through earthbound telescopes.

Of Saturn's 19 moons, only a few are prominent. Titan, the second largest moon in the solar system and the only one, besides Triton of Neptune, known to have an atmosphere, is usually visible as a faint star not far from the planet. Other fainter moons might be viewed too.

URANUS

Although there is very little that we can observe when we turn our telescopes to the solar system's three remote planets, they are interesting to discuss with children. Because they are so far away, their existence has been made known thanks to efforts of people with the drive and determination to carry out serious projects. In the cases of Uranus and Pluto, these projects involved observation of the sky for long periods of time without the slightest guarantee of success. For Neptune, two mathematicians had to fight for someone to take their calculations seriously enough to bother to search for the planet.

Only 8 years after England-based amateur astronomer William Herschel bought his first astronomy book, he was deeply involved in a systematic survey of the sky. The year was 1781; England had just lost the war with the 13 colonies in America. Observing with a telescope he designed and built himself, Herschel made notes on anything that was unusual—unusually colored stars, double stars, and fuzzy "nonstellar" objects. On March 13, he found a star in Gemini, which, he wrote, "appeared visibly larger than the rest...I suspected it to be a comet." But it was the strangest-looking comet anyone had seen—just a greenish blob of light, without any sign of the fuzziness and tail that often distinguish a comet. This object, whatever it was, was also moving much too slowly to be anywhere near the Earth. During that summer, with Gemini not visible in the night sky, astronomers wondered what this object could be. By the end of August, with Gemini rising in the morning sky and the new object visible once more, mathematician Anders Lexell confirmed what everyone was starting to suspect. This "comet's" orbit was almost circular, and never brought it closer to the Sun than 16 times the Earth's distance to the Sun. Herschel had not found a comet. Instead, he had made the first discovery of a planet in historic times.

Although Uranus is far away, a billion miles from the Sun, it is visible as a faint star through binoculars, and a telescope reveals the distant planet as a greenish round ball of light.

NEPTUNE

By the end of the first quarter of the nineteenth century, astronomers were starting to wonder about the strange orbit of the newly discovered planet Uranus. Like all planets, Uranus travels in a course that is affected mostly by the Sun, but to some extent by the other planets, especially Jupiter and Saturn. But all these forces of gravity did not explain the course in which the planet was moving. In 1834 John Couch Adams, a 23-year-old Cambridge University student, thought he had figured it out. Uranus's strange orbit, he concluded in 1841, could be explained if another large planet orbiting further out in the solar system was pulling on it. Adams even calculated where the new planet might be. He sent his prediction to his astronomy professor, John Challis, who in turn handed the work to George Airy, England's Astronomer Royal. Although both Challis and Airy seemed interested, neither did anything about beginning a search for a new planet at the position Adams provided.

It was 4 years before Urbain Jean Joseph Leverrier, a young French astronomer, also presented Airy with predictions for the positions of a new planet. Airy studied this paper and commented that Leverrier's predictions agreed within 1 degree (a thumb held against the sky is about 1 degree) of what Adams had predicted. Airy finally suggested that Challis search for the object. But two things went wrong: Airy forgot to tell Adams about Leverrier's work, and Challis, instead of going right to the predicted position, mounted a cumbersome star by star search over a large area of sky. Challis even observed the new planet twice without recognizing it as an interloper against the starry background.

Leverrier also tried to mount a search in his native France, but he met with little interest there. Then he went to Johann Galle at Germany's Berlin Observatory. Galle was definitely interested and began a search at Leverrier's predicted position. The very next night, he found an object that was not on their star charts and after a short while they noted that it was moving slowly. In a miracle of mathematics, a young French scientist had used the orbit of one planet to find another. When announced, the discovery caused a sensation and became such a *cause célèbre* that when George Airy attempted to associate Adams with it as well, a storm of controversy erupted. The French would have none of this, and the ensuing recriminations rocked the scientific establishments—and the public presses—of both nations. But neither Adams nor Leverrier were personally involved in the politics. After the tumult died down, they became close friends.

CLYDE TOMBAUGH AND THE DISCOVERY OF PLUTO

Clyde William Tombaugh was born on an Illinois farm in 1906. His early days on farms there and in Kansas ended with a hailstorm in June of 1928, which destroyed a crop as well as any enthusiasm he had for agriculture. In high school, Tombaugh became interested in what the geography was like on other planets. He built his first telescope, learned from his mistakes, and then made a very good 9-inch reflector. He used this telescope to sketch the planets. He decided to submit his drawings to the Lowell Observatory in Flagstaff, Arizona, to get their opinion of his work. Lowell's director, V. M. Slipher, was so pleased with the young amateur's work that he offered Tombaugh a observing job on the spot. When Tombaugh boarded the train at the start of 1929, he did not have enough money in his pocket for the return trip.

Shortly after Tombaugh arrived at Lowell, Slipher told him that his main project would be to carry on the search for a new planet that Percival Lowell had started in 1905 in hopes of repeating the Leverrier–Adams triumph with Neptune. Three searches had already been unsuccessful, but a new, wide-field telescope and a blink comparator would, they hoped, make the project more fruitful. Tombaugh quickly learned that with an optical device called a blink comparator, he could examine two photographic plates of the same part of the sky whose alternating views would allow him to see quickly if any object had moved in the sky between the time the two plates were taken. He also calculated that a trans-Neptunian planet should have a specific rate and direction of motion.

On February 18, 1930, Tombaugh was "blinking" two glass photographic plates he had taken, centered on the bright star Delta Geminorum. Just after four o'clock that snowy afternoon, he saw the two images of a faint object. "That's it!" he exulted. After checking the image with a third plate he had taken, he walked down the corridor to the director's office. Trying to keep as nonchalant as possible, but hardly able to control his excitement, he proclaimed: "Dr. Slipher, I have found your planet X."

Percival Lowell had mounted a search for a new planet based on the gravitational effect it would have on Uranus and Neptune. But the planet Tombaugh found had too little gravity to have any effect on these other worlds. It is a tribute to his observing skill and perseverance that Clyde Tombaugh discovered the ninth planet. After leaving Lowell in 1946, Tombaugh spent a decade at White Sands, New Mexico, where he developed the telescopes used to track the V2 rockets and other missiles that represented the dawn of the Space Age. He lived in Las Cruces, with a huge, home-made reflector telescope towering above his home in his backyard, and died in 1997.

ACTIVITY: SCALE MODELS OF THE PLANETS

Size Models: The Diameters of the Planets— a Two-Dimensional (2-D) Approach

BACKGROUND: Planetary scientists sometimes use the Earth as a reference point in making measurements. For example, the distance between the Earth and the Sun is called an astronomical unit or AU, and is used for distances in the solar system. Another example is Earth's atmospheric pressure (14.7 pounds per square inch), which is referred to as 1 atmosphere (1 atm), and the atmospheric pressure of the other planets is sometimes expressed in this unit. In the following activity, we use Earth's diameter as the base unit in constructing the diameters of the planets.

MATERIALS:

Paper, construction and butcher
   block
Scissors
Rulers
Compasses

String
Glue
Crayons or colored pencils
Calculators

OBJECTIVE: Students will construct an accurate scale model of the diameters of the planets and arrange them in the correct order from the Sun.

PROCEDURE:

1. Discuss what a scale model is. A unit of measurement on the model stands for a much larger unit on the object it represents. On any map, point out the legend and talk about what a centimeter (cm) or inch on the map represents in actual distance. Scale model cars and rockets, for example, can also be useful visual aids.

2. Introduce or review the terms diameter and radius. Although the planets are spheres, the models we are making will be flat. They represent the distance across the planet if it were cut open. You can cut an orange in half to illustrate this and show the diameter and radius.

3. Practice using the compass to make circles on scrap paper. Make sure the students understand that for a 6-cm circle, they must set their compass at 3 cm. For larger circles, demonstrate how to use two pencils and a string (it usually takes two students to do this).

4. Have the students make a chart of the diameters of the planets. Discuss the range in size from Pluto to Jupiter. Ask students to name the planets that are similar in size. They should also try to visualize how big each planet will be in comparison to the Earth. Explain that we will use Earth's diameter (12,756 km) as the base unit for the model. Using calculators, divide the diameter of each planet by 12,756. Round to the nearest tenth and fill in the data on the chart in column 3.

5. Our scale will be Earth's diameter equal to 3 cm. Multiply each number in the third column by 3. Fill in the chart.

6. Divide the numbers in the fourth column in half. This is the radius. Fill in the chart.

7. Students use the compasses or string to construct the planets.

8. Have pictures of the planets in the room so the students can color their models. Point out important features they should include. They then cut out their planets.

9. Arrange the planets in order from the Sun. Glue them on a long panel of butcher paper. Label the planets and write in the scale used in the model.

10. Have each group present its model to the class.
11. Emphasize that this is a model of diameters only. We cannot demonstrate the correct distances between the planets on the same scale. Also, in real life, the planets are not lined up in a row from the Sun.

EVALUATION: The teacher can evaluate the models for accuracy and neatness. Students should write about what they learned from making the model and discuss how their group worked to accomplish this task.

ADAPTIONS:

1. For younger students, use the patterns for the planets.
2. Have students calculate the diameter of the Sun on this scale and cut out a long panel of butcher paper to represent it. This is a good visual demonstration of how big the Sun really is compared to the planets. Then have them use calculators to divide the Sun's diameter by the diameter of each planet. This will tell the students how many of each planet could fit across the Sun. If desired, they can then build models by constructing and cutting out the necessary number of planets. Glue on to strips

MAKING A MODEL OF THE DIAMETERS OF THE PLANETS AND THE SUN
(Scale: One Earth Diameter = 3 cm)

Planet	Diameter (kilometers)	Diameter relative to Earth	To scale (× 3 cm)	Radius to scale (cm)
Sun	1,392,000			
Mercury	4,878			
Venus	12,104			
Earth	12,756	1.0	3.0	1.5
Mars	6,794			
Jupiter	142,984			
Saturn	120,536			
Uranus	51,118			
Neptune	49,528			
Pluto	2,302			

MAKING A MODEL OF THE DIAMETERS OF THE PLANETS AND THE SUN
(Scale: One Earth Diameter = 3 cm)
(Answer Key)

Planet	Diameter (km)	Diameter relative to Earth	To scale[1] (× 3 cm)	Radius[2] to scale (cm)
Sun	1,392,000	109.1	327.3	163.5
Mercury	4,878	0.4	1.2	0.6
Venus	12,104	0.9	2.7	1.4
Earth	12,756	1.0	3.0	1.5
Mars	6,794	0.5	1.5	0.8
Jupiter	142,984	11.2	33.6	16.8
Saturn	120,536	9.4	28.2	14.1
Uranus	51,118	4.0	12.0	6.0
Neptune	49,528	3.9	11.7	5.8
Pluto	2,302	0.2	0.6	0.3

[1]Also used for Options 1–3 on the following pages.
[2]Also used for Option 4 on following pages.

of butcher paper the same length as the panel representing the Sun's diameter (a lot of work!).
3. For a more challenging assignment, let the students decide on their own scale. It could be smaller or larger than 3 cm. Check their numbers before they begin to construct their models.

Size Models: The Diameters of the Planets— a Three-Dimensional (3-D) Approach

The following exercises refer to the table on page 182.

PLANETARY SIZES: The table on page 182 shows the names of the planets, their true diameters (in miles and kilometers), and their size relative to Earth. Column 5 gives a scale model of the planetary sizes relative to a 0.5-inch Earth (very close to 1:1,000,000,000). Column 6 gives the sizes of objects we have used in workshops to approximate the relative sizes of the planets (weather balloon for

Size Models: The Diameters of the Planets—a 3-D Approach
Planetary Sizes (Diameters) and Scale Models

Planet	True size		Scale size		Veggie and size		Weight, Pounds
	Miles	km	Earth = 1	in (mm)			
Sun	865,000	1,392,000	109.1	54.6 (1,385.6)	N/A	~55 inch	2,780
Mercury	3,031	4,878	0.38	0.19 (4.9)	Pea	5 mm	38
Venus	7,521	12,104	0.95	0.47 (12.1)	Small radish	0.5 inch	90
Earth	7,926	12,756	1.00	0.50 (12.7)	Mac nut	0.5 inch	100
Moon	2,160	3,476	0.27	0.14 (3.5)	Pea	3 mm	16
Mars	4,222	6,794	0.53	0.27 (6.8)	Small grape	6 mm	38
Jupiter	88,851	142,984	11.19	5.6 (142.2)	Lettuce	6 inch	287
Saturn	74,902	120,536	9.45	4.7 (120.1)	Cantaloupe	5 inch	132
Uranus	31,765	51,118	4.02	2.0 (51.1)	Lemon	2 inch	93
Neptune	30,777	49,528	3.85	1.9 (48.9)	Lime	2 inch	123
Pluto	1,430	2,302	0.18	0.09 (2.2)	Sunflower seed	2.5 mm	3

the Sun, styrofoam balls for the giant planets, and beads for the smaller planets). This is one-half the scale used for the paper models of the planets and is a convenient scale for balls, beads, and produce, while the paper model scale is suitable for using a compass for drawing all the planets. Column 7 is our Produce in Space model for the planets. For the smaller planets you can go healthy (nuts and seeds) or not (cinnamon candies, etc.). Column 8 gives the weight of a 100-pound student on the other planets, the Sun, and the Moon.

OPTION 1—SCALE A, EARTH = 3 CM (COLUMN 4):   Assemble a 3-D model solar system from household objects using the same scale. The teacher can provide the objects or can challenge the students to look for objects that match the given diameters. This option is a stepping stone to the volume-based scale model described in Option 2, i.e., allowing the students to explore ways of modeling other than 2-D diameter-based diagrams.

Hint: Start with a 12-inch (close to 33.6 cm) "disposable" aluminum pizza pan for Jupiter. Lids from large, round plastic cake or pie keepers can also be used for Jupiter. Try plastic lids from storage bowls or containers or round straw baskets for Saturn, Neptune, and Uranus. A half-dollar, fabric softener lid, or 16-oz. rubbing alcohol lid can represent Earth. A dime is slightly larger than, but very close to, the diameter of Mars; a food coloring bottle cap or $\frac{5}{8}$-inch button can be used. A quarter is just a bit smaller than, but close to, the size needed for Venus. The caps from 2-liter soda bottles, cough syrup, or a 24-ounce lemon juice bottle can be used. A $\frac{1}{2}$-inch button can represent Mercury. The hole from a standard paper punch, the eraser from the end of a pencil, the diameter of a plastic straw, or a $\frac{1}{4}$-inch button can be used for Pluto.

Allowing the students to provide the "planets" from home or to select the "planets" from a variety of teacher-provided objects provides opportunities both to practice metric measurements and to experience tangible examples or what 3 or 12 or 33 cm really looks like.

Other suggestions to try: shampoo lids, nuts, washers, plastic plant saucers, half-gallon ice cream lids, and embroidery hoops.

OPTION 2—SCALE B, EARTH = 1.5 CM (COLUMN 5):   It is difficult for both students and adults to visualize a 3-D size model when they are accustomed to discerning the relative sizes of planets based on 2-D diagrams in books or on posters. Using styrofoam balls, beads, and pompons or fruits and vegetables to create a scale model based on volume rather than diameter gives new insight into the relative sizes of the planets seen as spheres. *Remember*: The volume of a sphere is its diameter cubed. If the Earth's diameter is about 3.8 times the diameter of the Moon, Earth's volume is about 50:1 relative to the volume of the Moon!

Assemble a 3-D model solar system (one-half the size of the previous models) using fruits and vegetables (which allows for a reasonable size range of produce). Since it is more accurate to use a tape measure to measure the circumference of limes, tomatoes, etc., the teacher (or older students) will need to convert the given diameters to circumferences. To calculate the circumference of the produce needed, multiply the given diameter by 3.14 (π).

Hint: Start with a head of lettuce or honeydew melon for Jupiter and a cantaloupe for Saturn. Try small apples, tomatoes, or limes for Neptune and Uranus. Try small radishes or macadamia nuts for Earth, Venus, and Mars. Peas or seeds and nuts from trail mix work well for the very small planets. Halves of seeds can represent the asteroids. You can also substitute small, round candies for Pluto and Mercury.

Distance Model: Distances in the Solar System

PLANETARY DISTANCES: The table on page 185 gives the true (mean) distances of the planets from the Sun (in millions of miles and kms) and their distances in AU, where 1 AU is the mean distance of Earth from the Sun. Column 5 gives the distances of the planets (from the Sun) in solar diameters (107 suns laid side by side would be needed to stretch from the Sun to Earth). Column 6

Planetary (Solar) Distances

| Planet | Distance | | | | Distance in light-minutes | Period |
	Miles (millions)	km (millions)	AU	Sun = 1		
Mercury	36.0	57.9	0.387	42	3.2	87.97 days
Venus	67.2	108.2	0.723	78	6.0	224.54 days
Earth	93.0	149.6	1.000	107	8.3	365.24 days
Moon	0.238	0.383	0.003	30[a]	1.3[b]	27.32 days
Mars	141.7	228.0	1.524	164	12.7	686.98 days
Jupiter	483.7	778.4	5.203	559	43.3	11.86 years
Saturn	885.2	1,424.6	9.523	1025	79.5	29.46 years
Uranus	1,785.5	2,873.5	19.208	2061	159.7	84.01 years
Neptune	2,796.8	4,501.0	30.087	3230	250.4	164.79 years
Pluto	3,694.6	5,945.9	39.746	4285	332.1	248.5 years

[a]Relative to diameter of Earth.
[b]Light-seconds

39. *Solar system size model, shown by Paul Scalise.*

gives the time it takes light to travel from the Sun to planets
(186,000 miles or 299,800 km a second). Column 7 gives the orbital
period (year) for each planet. For the Moon these are all relative to
Earth.

PLANETARY SCALE MODEL DISTANCES: This scale model is good
for teaching students how to read maps. This is essentially at the
same scale as in the first table of the planetary sizes, except that it
is scaled to exactly 1:1,000,000,000 (1 meter = 1,000,000 km). Have

the students mark on a map the streets as listed in the "location" column or have them fill in the location column with streets from a city map starting from any other location such as your school. Streets in Tucson, Arizona, are given as examples.

In this table, there are three scales: the same as the previous table (rounded off) and $\frac{1}{100}$ that scale. The final column gives the time it would take for a spaceship to travel from the Sun directly to the planet at 50,000 miles an hour (about the speed of the Voyager spacecraft). You should understand that this is an oversimplification since spacecraft do not fly directly to planets; they essentially go into an orbit around the Sun that brings them close to the planet. In recent times, such as in the case of the Galileo, which is on its way to Jupiter, the spacecraft has had to go past other planets in order to get up enough speed to get it out to the giant planets. Thus, it may take several years for a spacecraft to get to these planets rather than a year or less.

The second scale is probably the best for doing a scale model of the solar system on a playground or field without getting students too far away from the Sun. Distances can be measured by pacing off, with a meter stick, with a tape measure, or with pre-measured colored cord. Colored macramé cord can be purchased at any crafts store for under $15 and should last for years.

On this scale, Alpha Centauri, one of the closest stars (other than the Sun), is 408 km away and Sirius, the brightest star (other than the Sun), is 848 km away.

Fruits or Produce in Space

Though this activity is meant for elementary grades, we have also used it for students in middle school.

1. Ask the students if they know what a model is. Ask them if they have ever had a model plane or car: it looks like the real thing, but smaller. Say that since the Sun and planets are too big to fit in the classroom, we will be making a model of the planets in the solar system and that we will be using fruits, vegetables, nuts, etc.

Planetary (Solar) Scale Model Distances[a]

Planet	Scaled distance			Location example: Tucson, AZ
	Feet	Miles	Meters	
Sun	0.0	0.0	0.0	Speedway and Campbell
Mercury	190.1	0.04	57.9	Plaza Hotel
Venus	354.9	0.07	108.2	Plaza Hotel
Earth	490.8	0.09	149.6	Norris
Moon	1.3		0.4	
Mars	748.2	0.14	228.0	Olsen
Jupiter	2,553.9	0.48	778.4	Norton (Tucson)
Saturn	4,673.9	0.88	1,424.6	Bentley (Country Club)
Uranus	9,427.4	1.77	2,873.5	Dodge (Alvernon)
Neptune	14,767.1	2.78	4,501.0	Belvedere (Swan)
Pluto	19,507.5	3.69	5,945.9	Beverly (Craycroft)

Planet	Scaled Distance[b]		Scaled Distance (1/100)[c]		Time (@50,000 miles an hour)
	Feet	Meters	Feet	Meters	
Sun	0.0	0.0	0.0	0.0	
Mercury	190	58	1.9	0.6	30 days
Venus	35	108	3.5	1.1	56 days
Earth	491	150	4.9	1.5	77.5 days
Moon	1.3	0.4	0.1	0.004	0.2 days
Mars	748	228	7.5	2.3	118 days
Jupiter	2,554	778	25.5	7.8	1.1 years
Saturn	4,674	1425	46.7	14.2	2.0 years
Uranus	9,427	2874	94.3	28.7	4.1 years
Neptune	14,767	4501	147.7	45.0	6.4 years
Pluto	19,508	5946	195.1	59.5	8.5 years

[a]Scale factor: 1 meter = 1,000,000 km (1:1,000,000,000).
[b]Sun diameter = 55 in (use original Sun diameter from Planetary Sizes table on p. 182) 1 meter = 1,000,000 km (1:1,000,000,000).
[c]Sun diameter = 0.5 in (use original Earth sphere from Planetary Sizes table) 1 meter = 100,000,000 km (1:1,000,000,000,000).

2. We then have the students hold each of the "planets" ordering them by size starting with Jupiter (the Sun will be 4–5 feet in diameter, about the height of a student). Then we reorder them by distance from the Sun.
3. We then go outside and make a scale model of the solar system, to the same scale as the planets. The students soon find out that we cannot do the whole solar system, so we must rescale the sizes.
4. By using "Jupiter" to represent the Sun (rescaling by a factor of 10) the solar system can be done on a typical playground. We have a child represent the Sun, Mercury, etc., each holding a sign with the name of the planet. One child is a spaceship going from one planet to the next. To give the students a feel for the distances involved, they are either told or they determine on their own what grade they would be in as they reach each planet.
5. This can be expanded in several ways. The original scale can be used with a map of the town around their school for determining the distances to the planets, allowing for map reading. Also, one can go beyond the solar system to measure the distances to the nearest stars. For this, we use airline tickets to "fly" the students to the appropriate world destinations, thus bringing in geography.

CONCLUSION

Teaching children about other worlds is important because it helps them to understand our own world better. Venus gives us an example of how our Earth might be if the greenhouse effect were to run out of control, and Mars shows us how life as we understand it could not exist in the thin atmosphere. At the opening of this chapter we made light in an entertaining way of the harsh conditions of these other worlds—as a way of impressing on young minds how fortunate we are to have our Earth and how fragile its environment is.

Chapter 10

Beauty and Danger: Comets, Asteroids, and Meteors

Although the solar system is best known as the Sun and its family of nine planets and their moons, the smaller bodies—the comets and asteroids—add a new dimension to its size, history, and beauty.

We do not know how many asteroids or comets there are. More than 6,000 asteroids have been given official numbers, and most of these have been given names as well. Since most asteroids orbit the Sun in a belt between Mars and Jupiter, they can be observed frequently and their orbits can be accurately determined. At least another 5,000 have been observed less frequently. Finally, more than 200 asteroids have been discovered whose orbits around the Sun, or paths, take them near or across the orbit of the Earth.

Since most comets travel in large orbits that loop across the solar system, they have not been tracked as carefully as the numbered asteroids. We know of almost 1,000 comets that have appeared since 1059 B.C., when a bright comet appeared during a war between two long-forgotten Chinese kings.

Meteoroids are the detritus of comets and asteroids, the small pea- or sand-grain-sized particles that comets and some asteroids leave in their wakes. When a meteoroid rushes through Earth's upper atmosphere, it heats the gases to incandescence, causing them to glow. We call this bright event a meteor.

COMETS: ADDING DRAMA TO HISTORY

In addition to their beauty, comets are exciting because they connect so readily to other important areas of learning, particularly history. A comet that appeared just two months after Julius Caesar's death was interpreted as symbolizing the fallen leader's journey to the stars. In 1181 a comet appeared to presage the death of Pope Alexander III, and only 17 years later another announced the passing of Richard I of England.

Since Comet Halley has appeared so regularly and so brightly at 76-year intervals, it has been connected to several historical events. In April 1066, the magnificent comet timed its visit to coincide with the Norman conquest, and its legacy became part of the Bayeux Tapestry, complete with its battle cry *"Nova stella, novus rex"* ("New star, new king")

Our slowly growing understanding of the nature of comets changed during the Renaissance. It was Edmond Halley who proved the extraterrestrial nature of comets beyond any doubt by his brilliant work on the comet that now bears his name. It took more than a spark of genius to note that among all the comets that have appeared since records have been kept, a single one seemed to appear at 76-year intervals.

In the last ten years, our understanding about what comets are and what their passages mean has reached a new level. In 1986, Comet Halley was studied extensively by telescopes all over the world and by a flotilla of spacecraft, which revealed that Halley contains particles of carbon, hydrogen, oxygen, and nitrogen. C, H, O, and N form the simple alphabet of life, and they exist in the comet in very nearly the same proportions in which they appear in every form of life on Earth. It is possible that in Earth's early history, a bombardment of comets might have deposited the organic materials and water that later took hold as the beginnings of life on Earth.

The second major leap occurred in 1994, when some 21 fragments of Comet Shoemaker–Levy 9 plowed into Jupiter at almost 140,000 miles an hour. From the behavior of this comet on its way to Jupiter and when it hit the planet, we are learning much about comets, planets, and the role of cometary impacts as a major tool in building the solar system (see chapter 11).

What Is a Comet?

When far from the Sun, a comet is a "dirty snowball" comprised of a nucleus of rocky material loosely held together by a glue of frozen gases. The nucleus is the source of all cometary phenomena. This tiny member of the solar system, almost never directly observed, generates some of the largest phenomena (comet tails) and the smallest objects (dust particles and free molecules and atoms) that have been seen in the solar system.

During its passage through the inner part of the solar system (the area from the Sun to Mars), the Sun's heat causes the ices to sublimate (change from solid to gas directly, without changing to a liquid first). Each time it passes the Sun, Comet Halley's nucleus loses the equivalent of about 1 meter's thickness of material.

Formed when a comet approaches the Sun, the coma is a halo of material surrounding the nucleus. Gas erupts from vents in the comet, carrying dust particles with it into this tenuous cometary atmosphere. The coma and nucleus together are referred to as the head of the comet. Surrounding the head is a huge cloud of atomic hydrogen gas. Depending on the size and activity of the comet and its distance from the Sun, this hydrogen envelope can be 1 to 10 million km in diameter.

Most comets have two-part tails. The ion tail consists of molecules released by the nucleus that have been ionized (forced to lose an electron and acquire a positive charge) by solar ultraviolet and X-radiation. This tail is bluish in color photographs. The dust tail is generated by a different process. Dust particles about 1 micrometer in size are carried into the coma by the "wind" of molecules released as the nucleus sublimates. Solar radiation pressure affects dust particles the way wind drives a sailboat. The dust is pushed away from the Sun, but it also moves along with the comet. The result is a curved dust trail.

Where Do Comets Come From?

During the centuries in which comets have been observed, surprisingly little has been learned about their origin, evolution, and the processes occurring within them. Most of our detailed knowledge of comets has been acquired in the past few decades, and even this is incomplete. The image of a long, flaming tail and a bright, starlike head is dramatic but hardly accurate for most comets. It is infrequent that we have a comet of such magnitude, and rarely do we see one of truly gigantic proportions.

Some comets are believed to originate in the Oort Cloud, a huge collection of remnant cometary nuclei dating from the formation of the solar system. This cloud of Sun-orbiting comets extends from perhaps 20,000 to beyond 150,000 AU. (1 AU is the average distance between Earth and Sun, which is 150,000,000 km, or 93,000,000 miles.) There are no perceptible boundaries to this spherical shell and the billions of comets there are so widely scattered that the word "cloud" is misleading. When a comet's course is changed, perhaps by a passing star, it will start on a millions-of-years-long journey to the inner solar system. Astronomers have now found about 100 objects that orbit in the plane of the solar system between the orbit of Uranus to well beyond the orbit of Pluto. These are thought to be part of the "Kuiper Belt" of frozen cometary nuclei that are remnants of the material that remained after the planets were formed.

Sungrazing Comets

In a set of solar system objects noted for character, the sungrazers certainly hold their own. Their distinguishing characteristic is that they come extremely close to the Sun, sometimes less than a million miles from its photosphere. The Kreutz group of sungrazers probably began with the breakup of one great comet possibly more than 10,000 years ago. Included in this group are the Great Comet of 1882, Comet Pereyra of 1963, Comet Ikeya–Seki of 1965, Comet White-Ortiz-Bolleli of 1970, and a number of comets found by the orbiting Solwind and Solar Maximum Mission spacecraft. Both the 1882 and the 1965 comets rounded the Sun at times that were favorable for splendid shows. These two comets almost certainly split from a single comet that rounded the Sun in 1106.

Comet Hunting

Over at least two centuries, the saga of comet hunting as a dramatic sport has captured the public imagination and certainly gets lots of press coverage whenever a bright or interesting comet appears. Because comet hunters compete for the holy grail of discovering the 15 or so new comets that appear each year, their activities amount to a single life-long game.

Early in the nineteenth century, Jean-Louis Pons began his observing career as a doorkeeper at an observatory in France. After some success as a discoverer of comets he was promoted to a position of observatory astronomer, and his career total is at least 27 comet finds. Leslie Peltier, an amateur astronomer from Delphos, Ohio, began his comet hunt in 1922 with a borrowed 6-inch refractor. When he learned that his new instrument had an honored past, with three finds early in the twentieth century by Zaccheus Daniel, he decided to continue its tradition. Three years later, on November 13, 1925, he discovered his first comet, and over the next three decades he found 11 more, one of which became easily visible to the naked eye.

Kaoru Ikeya of Japan represents another success story. As a worker in a piano factory, he had to support his family, whose name his father had disgraced through a business failure. He found his first comet on January 2, 1963, and his second came some 18 months later. On September 18, 1965, Ikeya discovered an eighth magnitude patch of haze in Hydra, an object that Tsutomu Seki discovered independently only 1 hour later. Shortly after the discovery of Comet Ikeya–Seki was announced, an orbit was calculated that projected the object to approach within 300,000 miles of the Sun's surface, resulting in the most spectacular comet show in more than half a century.

William A. Bradfield discovered his first comet in 1972 and his second in 1974, just after the passage around the Sun of the well-publicized Comet Kohoutek. Although it was more favorably placed for observation than Kohoutek, the second Comet Bradfield, at least as far as the public was concerned, seemed lost in Kohoutek's publicity. In the next year Bradfield discovered two more, and two more again in 1976. He found his 16th in 1992.

By discovering his first comet in the spring of 1978 with a 16-inch-diameter reflector, Rolf Meier became the first successful Canadian comet hunter. In 1979 he discovered a faint twelfth-magnitude comet, one of the faintest visual discoveries ever, an achievement exceeded in 1984 when he found one even fainter.

The Shoemaker–Levy Comets

Around 1982, Eugene and Carolyn Shoemaker began a husband-and-wife program of searching for asteroids. Using Palomar Observatory's first telescope, the 18-inch-diameter Schmidt camera built in 1936, they took a large number of photographs of the sky during the course of a seven-night observing period held each month. Although their purpose was mostly to locate asteroids that could be in courses that cross Earth's orbit, they began finding new comets in 1983. By 1994, using a stereomicroscope designed by Gene, Carolyn had found 32 new comets. Thus Carolyn Shoemaker has the all-time record for finding comets. In recent years Henry Holt has accompanied the team, and there are several Shoemaker–Holt comets. By 1989 David Levy began observing with them, and together they have found a long series of comets. As all comets found before the end of 1994, the Shoemaker–Levy comets were designated in two ways. The comets that come round once and return rarely or never are simply named Shoemaker–Levy. The others, called periodic comets, return every few years and are given numbers. In April 1992, for example, the periodic comet Shoemaker–Levy 8 appeared. Shoemaker–Levy 9, the comet that struck Jupiter, is the subject of chapter 12.

Can a teacher, parent, or child discover a comet? In 1968, Mark Whitaker, a high school student, found Comet Whitaker–Thomas after searching for only three nights. However, such an event is very rare. We tell the stories of discovery for background information, and in the hope that some readers might actually decide to begin a search for these elusive objects.

Activity: Building a Comet

PURPOSE: To show how a comet is made and how it works.

MATERIALS:
For each comet:

2 cups water	2 spoonfuls sand or dirt
2 cups crushed dry ice (wt. should be 2 pounds)	A dash of ammonia
	A dash of organic material, e.g., dark corn syrup

Other materials:

An ice chest	A hammer, meat pounder, or rubber mallet
1 large plastic mixing bowl or small plastic bucket per comet	1 sturdy plastic mixing spoon per comet
2 medium-sized plastic garbage bags per comet	Paper towels and/or newspaper

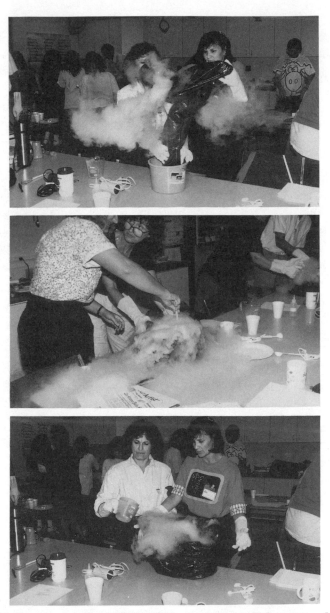

40. *Making a comet. ARTIST 1993: (top) Betty Fulcomer and Elsa Fimbres, (middle) Kathy Eichhorst and Donna Stefanek, (bottom) Betty Fulcomer and Elsa Fimbres.*

Work gloves, or heavy plastic
   gloves

Flashlights
Hair dryers
Small magnifying glass (optional)

PROCEDURE:

1. Protect your work surface with newspaper and have plenty of paper towels or sponges for cleanup.
2. Crush the dry ice with the hammer or mallet into powder. Leave some chunks about the size of walnuts.
3. Double the plastic bags and place inside the bowl.
4. The person who will be handling the dry ice *must* wear protective gloves. Both partners should wear gloves, if they are available.
5. Put two cups of dry ice into the bags inside the bowl. Add the dirt, ammonia, and corn syrup. Mix. [Option: Mix the dirt, ammonia, and corn syrup into a small container (paper cup, aluminum pie can, etc.) first and then add the dry ice.]
6. Add the water. A reaction will occur immediately, producing a cloud of vapor.
7. One partner should hold the bag shut at the top, while the gloved partner squeezes the bottom of the bag to form a ball of ice. It does take a good deal of strength to get the dry ice, sand, and water mix to form a ball.
8. If the ingredients do not form a ball, try adding a small amount of water.
9. Once there is a solid ice ball within the bag, turn it out onto a sturdy paper plate or bowl.
10. Look for evidence of sublimation, i.e., of the solid turning into a gas. A small magnifying glass might be helpful here.
11. Hold a hair dryer a few feet from the comet and turn it on to create a wind on one side of the comet. This wind simulates the pressure of radiation from the Sun on a comet. Shine the flashlight on the opposite side of the comet to illuminate the tail. The gas (or vapor) forming the tail is the result of dry ice sublimating. The white vapor you see is water ice frozen on bubbles of cold carbon dioxide, an invisible gas.

12. *Caution*: Do not handle the dry ice without protective gloves. Enforce safety rules with the students. It may take several hours for the unused dry ice to sublimate completely. Be sure that the dry ice is in a safe place where students will not be tempted to handle it.

ASTEROIDS

Asteroids are not as dramatic as comets, so they get less attention in classrooms. But they are an important part of our solar system. Discovered in 1801, the first asteroid, only 700 miles wide, started a mystery: Why was the new planet so tiny? Within 2 years three others were found. Had a larger planet broken apart long ago? Scientists now suspect that the asteroids represent material that never did gather to form a planet, probably because of constant disruptions by Jupiter's gravity.

Most textbooks claim that the asteroids all orbit in a belt between Mars and Jupiter. For most of the asteroids, this statement is true. But the ones that do not orbit there are very important: there are believed to be some 2,000 rocky bodies more than 1 kilometer (about two-thirds of a mile) across that actually cross the orbit of Earth. Anyone who has seen Meteor Crater near Flagstaff, in northern Arizona, knows that sometimes these asteroids actually collide with the Earth with devastating results.

What Is the Difference between a Comet and an Asteroid?

The ancient definitions of comet and asteroid still work today in some respects. Through a telescope, a comet looks like the fuzzy star or "hairy star" that the Greek derivation of its name implies. An asteroid looks like a star through a telescope, and on a long exposure photograph it looks like a moving streak. But the solar system is a continuum of objects, and when a comet is not active, with its coma and tail of dust and gas, it can look exactly like an

41. *Finished comet by Mary DeStefano, ARTIST 1993.*

asteroid. Recently, three asteroids have been found that later displayed the comas and tails that define comets. Comet Shoemaker–Levy 2 was discovered on films taken by the Shoemakers and David Levy on two photographic films as an asteroidal trail that was called 1990 UL3. When Steve Larson and Levy used a much larger telescope and a sensitive computerized detector called a CCD (charge-coupled device), they found that the object sported a tail at least 40,000 miles long, and so the object was renamed in the tradition of comets. As asteroid 2060 Chiron, a large object out beyond Saturn that is at least 100 miles wide, traveled closer to the

Sun in the years following its discovery, it developed a coma and a tail and now appears more cometlike than asteroidal.

What Asteroids Are Made Of

By studying asteroids with telescopes scientists have learned much in the last few decades about the likely composition of asteroids. Scientists have also been able to study actual pieces of asteroids—meteorites. On its way to Jupiter, the Galileo spacecraft flew by two asteroids. Pictures of these asteroids show that the asteroid surfaces are covered with craters, the remains of impacts similar to what we see on the Moon. We now know that asteroids are largely made of rocky material and iron. However, many of them appear to contain dark material (possibly organic material) and water. Most asteroids appear to have changed very little since the solar system formed. Some of them have surfaces similar to volcanic rocks on Earth. Others seem to be made up entirely of metal, perhaps the cores of larger asteroids that have been broken up by numerous collisions with other asteroids. One of this book's authors, Larry Lebofsky, discovered water on the largest asteroid 1 Ceres in 1978. Since then he and his colleagues have found water on nearly a third of the asteroids. This is not likely to be ice on the surface of the asteroid, but is probably water bound to the rocks themselves, similar to Earth-like clay material. This shows how similar asteroids and comets really are!

METEORS

While most children will not relate too well to the complex physics involved when a sand-grain-sized body races through the atmosphere, they do relate to meteors. Few children have ever seen an asteroid, but many have made their wish upon a falling star. Falling stars are not stars, but meteors.

The first point to be made is that a meteor is not a star that actually falls from the sky. After each meteor falls there are still just as many visible stars as there were before. Every star is a sun, in some ways like our own. These distant suns will not fall to Earth. A meteor is an *event* that takes place pretty close to us, perhaps 40 miles above in the Earth's own upper atmosphere.

Although meteors can be seen on any night of the year, at certain times of the year the Earth rushes through a stream of particles that are strung along the orbit of a comet. The most famous of these streams is the Perseids, which peak on August 12 and are so named because their meteors seem to radiate from the constellation of Perseus; we say that the shower's *radiant* is in Perseus. Observers watching at this time might also see some meteors from the Delta Aquarid shower, which hits its maximum on July 29 with meteors one can trace back to a point in the sky near the star Delta in the constellation of Aquarius. During the school year, the Earth encounters several streams to produce what we call meteor showers. The Orionids, consisting of dust from Halley's comet, occur around October 20, and the Taurids, including some very dramatic bright meteors, are visible during the first part of November. The Leonids occur on November 17, and the Geminids reach their strong maximum on December 13. Kicking off the new year are the Quadrantids, which are active for only a few hours on January 3. The Lyrids peak at the end of April, and the Eta Aquarids, also from Halley's comet, are active around May 4.

We see that meteors in a shower appear to radiate from a single point; this is due to perspective. The effect is similar to looking down a railroad track. The tracks appear to converge far away, but as they are parallel, obviously they can't. As the meteoroid particles encounter Earth, they all hit on parallel paths, but owing to the effect of perspective, we can trace them back to a single point.

METEORITES

On April 26, 1803, a shower of thousands of small stones fell over northwestern France, terrifying many people. On November 30, 1954, an Alabama woman was inside her home when a small meteorite shattered her roof, bounced off her radio, and proceeded to hit her. People all over the midwestern United States witnessed a fireball—a very bright meteor—as bright as a quarter Moon on the evening of Friday, October 9, 1992. Lighting up the sky, it took some 15 seconds to cross the sky and disappear in the northwest. Meanwhile, Michelle Knapp, a high school senior spending the evening in her Peekskill, New York, home, was startled by a crash outside. She went out to find her car's trunk crushed and a football-sized rock sitting next to it. Angered by what she thought was an act of vandals, she called the police, who in turn summoned some scientific help. It turned out that the boulder was the fireball that so many had seen, ending its journey ignominiously by hitting a car.

Almost all of the known meteorites are pieces blasted off asteroids. However, we now have discovered several meteorites that have come from the Moon or Mars. While it is difficult for an inexperienced person (and many times for even an expert) to tell if a rock is a meteorite, some are more easily recognizable than others. Iron meteorites, made of iron and nickel, are much denser than typical Earth rocks. Their fusion crust, formed by the heat of their passage through the Earth's atmosphere, is usually brownish in color. For this reason it is usually confused with magnetite. However, a true iron meteorite has a metallic silver-colored interior. Many stony-iron meteorites may also be easily identifiable, but are much rarer. The most common meteorites are the stony meteorites, and to the untrained eye they look just like ordinary Earth rocks. Many of them contain condrules, spheres of previously melted rocky material that date back to when the asteroids were formed.

Chapter 11

Advanced Project: Observing Meteors

If you can get a group of children together near the time of one of the major annual meteor showers (see chapter 10), getting them to observe it is a learning experience that can be a lot of fun.

PURPOSE:

1. To introduce the children to the night sky, its stars, its constellations, and its planets.
2. To demonstrate how the Earth picks up particles as it moves through space.
3. To demonstrate how scientific observing sessions are conducted and recorded.

MATERIALS AND CONDITIONS:

Lawn chairs if the group is small or blankets if it is large

A flashlight with a red filter, used for keeping your eyes adapted to the darkness while you record observations. The purpose of the filter is to allow recording of each meteor without losing the eye's adaptation to the low level of brightness of the night sky.

Recording materials, including paper recording forms and pencil. The paper should be arranged thus:

Meteor Record

Date:
Place:
Observer:
Time Begun:
Breaks from   to
Time Ended:

Meteor Number	Brightness	Shower
1		
2		
3		
4		
5		

A clear night around the time of one of the year's major meteor showers. During the school year, the best times are:

From December 8 to 13: Geminids

End of April into early May: Lyrids and Eta Aquarids

First two weeks of August: Delta Aquarids and Perseids

PROCEDURE:

Preparation: Although the main reason for observing the sky on the night of a strong showing of meteors is for educational value, the children might be impressed to learn that their observations might have real scientific value if they are conducted conscientiously. The observing session has the most value if it occurs near the end of a unit on comets and meteors, but the fact that meteors are best seen on a few nights of the year means that the session will probably take place out of the children's learning sequence. In any event, they should be aware of what meteors are, why they appear to trace back to a single point in the sky, and why they might come from more than one point during the night's observing if a second radiant is active (see page 203 for the discussion of radiant). It is important to point out that the meteors will not all begin at the radiant. They could begin anywhere in the sky and end anywhere. The idea is to trace their path *backward*, in the direction of the radiant. Doing this with several meteors should pinpoint the radiant.

Learning about Constellations: Since the meteors appear in different parts of the sky, this is a wonderful opportunity to teach the children about constellations (see chapter 3). You should tell them something of the mythology and history of the constellation in which the shower's radiant is found as well as stories about other nearby constellations. For example, the story of Perseus includes the woes of the princess Andromeda, whose rock-bound trials stem from the arrogance and vanity of her mother, Cassiopeia, who with her father, Cepheus, are represented as nearby constellations. Perseus rescued Andromeda and brought her to the sky.

Learning about Star Magnitudes: Since observers will record a meteor's brightness, this is a good time to introduce them to the concept of magnitudes. They should easily understand that the stars are of different brightnesses (see chapter 13), and it should also be easy to explain why this is so by using streetlights as an analogy. The streetlight near your house appears brighter than one a few houses down because it is closer; similarly nearby stars appear brighter than more distant ones. However, big lights at a distant shopping mall may appear brighter than the nearby lights because they really do have brighter bulbs. Similarly, luminous stars are bright, even though they might be far away.

For recording the magnitudes of meteors, however, all the children need to know is how we measure brightnesses. Here is a list of stars of particular magnitudes (see the names on the figures in chapter 3):

−2	Sirius
0	Vega · Antares · Arcturus · Capella ·
1	Deneb · Spica · Altair · Aldebaran · Pollux ·
2	Polaris · Alpheratz · Dubhe ·
3	Mebsuta (Epsilon Geminorum) · Gamma Boötis · Gamma Ursae Minoris · Gamma Virginis ·

OBSERVING:
1. Arrange the children in groups of four or five.
2. Have each child face a particular direction—north, south, east, or west—in the sky. A fifth child can scan the overhead region.
3. Have each child record the start time on the observing form.
4. For each sighting of a meteor, the child should first estimate its brightness by comparing it to appropriate stars listed above. A meteor that is as bright as Mizar, the middle star of the Big Dipper's handle, would be assigned magnitude 2.
5. For each meteor, the child should then trace the path backward in the sky. Does it appear to originate from the radiant? If it does, then the child should record the shower's name as P for Perseid, D for Delta Aquarid, and so on. If the path goes backward to a part of the sky nowhere near the radiant, the child should write N to identify a nonshower meteor. In reality, almost all meteors come from showers, and in August no fewer than six shower radiants are active. We note only the major ones for simplicity.
6. After a period of time for observing—perhaps an hour—have each child record the ending time on the observing form.

CLOSURE: There should be a lot of free discussion about what the children learned from their observing experience, for it could be a rare learning privilege that they will remember for the rest of their lives. Each meteor they saw was a miniature impact of a particle against the Earth's atmosphere. These tiny objects hit often, as we found out by watching them. But the bigger they get, the less often they appear. The children might have been lucky enough to see a very bright meteor. If they did, they know that it was a relatively rare occurrence during the course of the night. In fact, a simple study of the magnitude distribution of the children's own observations should show that the fainter meteors appeared more often than brighter ones.

The larger a particle is, the less frequently one of that size is seen to fall to Earth, and the brightest meteors the children saw were probably from objects no bigger than grapes. Such a small object will burn up in the Earth's atmosphere before reaching the ground. On that scale, an object the size of child's desk should hit somewhere over the earth every week (usually over some inaccessible part of Earth's oceans) and something as big as a room every year. An object 1 km across (some 2/3 of a mile) hits every 100,000 years and can cause severe damage. As we've noted before, 65 million years ago a comet or an asteroid at least 10 km in diameter (about 7 miles) led to the extinction of most forms of life on Earth, which is an event that might occur on an average of every 100 million years. Thus, on this special night, the children got a chance to observe a part of a major process that affects the Earth and its inhabitants in serious and exciting ways.

Chapter 12

A Week to Remember: Comet Shoemaker–Levy 9 and Jupiter

July 16, 1994, was a new kind of day for astronomy. For the first time ever, the world's entire arsenal of large telescopes was pointing at a single object in the sky. The planet Jupiter was about to be bombarded by a string of 21 comets. For the first time in history, we would see the solar system in action as a comet ended its life in the dense atmosphere of mighty Jupiter. We tell the story of the comet crashes in this book to give parents, teachers, and children a taste of what happened during this exciting time.

WHY IS A COMET IMPACT SO IMPORTANT?

As we've pointed out before, some 65 million years ago, a large comet (or possibly an asteroid) smashed into the coast of what is now the Yucatan Peninsula in eastern Mexico. As tons of debris shot back into the sky and fell again all over the world, the temperature of the air grew very hot. Soon a thick cloud of debris entombed the Earth for several months. Since the Sun couldn't shine though, photosynthesis stopped, plants died, and animals starved to death. Before it was all over, more than 70 percent of all the species of life-forms became extinct, including all the dinosaurs.

In recent years we have gathered considerable evidence to show that an impact actually did take place on Earth 65 million years ago. The most powerful piece of evidence—the smoking gun—is a large crater more than 100 miles wide.

Other comet impacts helped to shape the course of life on Earth. In fact, early in the Earth's history comets might have been the source of Earth's supply of organic materials—carbon, hydrogen, nitrogen, and oxygen—and consequently the source of its water as well.

WAITING FOR THE IMPACTS

The armada of telescopes watching the S–L 9 crash included two important instruments not on Earth—the Hubble Space Telescope (HST) and the Galileo spacecraft. The timing was perfect for both telescopes. Just 6 months earlier, an extraordinarily successful repair mission had turned HST into the most powerful astronomical instrument ever turned toward the sky, and a spacecraft named Galileo, 19 months away from its encounter with Jupiter, had a vision of Jupiter we never get on Earth. From Galileo's viewpoint Jupiter looked like a gibbous moon. The craft's 10-cm (4-inch) diameter telescope saw a small part of the night side of the giant planet. It was just our luck that the comets were hitting Jupiter in that small sliver of darkness. We would not see any of the direct hits from Earth. But Galileo's eye saw the slashes as each fragment fell through Jupiter's upper atmosphere.

HOW COMET SHOEMAKER–LEVY 9 WAS DISCOVERED

On March 25, 1993, Gene and Carolyn Shoemaker and David Levy discovered a comet that was on a collision course with

42. *As fragment W disintegrated in Jupiter's atmosphere, the Galileo Space Probe was watching (NASA/JPL image).*

Jupiter. Following a naming tradition that has gone on for more than 200 years, this comet was named Shoemaker–Levy 9 after its discoverers; it was the ninth find in a series of comets that travel around the Sun in short-period orbits. The team had five other comets, all named Shoemaker–Levy, that travel in long, wide loops. These comets won't return for many thousands of years, if ever.

The new object really did look like a comet that someone had stepped on. Instead of a single coma and tail, it had a bar of coma, with a series of tails stretching to the north. But the strangest part of it was that on either side of the bar was a pencil-thin line.

Three things needed to be done. First the discovery needed to be confirmed. Jim Scotti, who was observing that night from the Spacewatch telescope atop Kitt Peak, Arizona, agreed to photograph the comet. He was stunned by the image he saw. "There are at least five discrete comet nuclei side by side," Scotti said as he described the view through his telescope, "but comet material exists between them; I suspect that there are more nuclei that I'll see when the sky clears." He also observed two large wings of dust that made the comet look a little like a B-2 bomber.

The second order of business was to report the discovery to Brian Marsden, director of the International Astronomical Union's Central Bureau for Astronomical Telegrams. The following day, Marsden's office issued a circular announcing the new comet. So unusual was the object's appearance that observers around the world began to observe it immediately. Using the University of Hawaii's 88-inch reflector, Jane Luu and David Jewitt obtained a magnificent image of the comet. They later resolved 21 separate subnuclei strung out, they wrote, like pearls on a string.

JUPITER TEARS A COMET TO PIECES

In the months after discovery, the comet provided astronomers with a series of surprises. The first was that it apparently was disrupted because of some previous close approach to Jupiter. This was a tidal effect: Anyone who lives near an ocean sees how the gravitational pull of the Moon, and to a smaller extent the Sun, causes a gentle surge of water from a low level to a higher level and back again. We call these surges the tides.

This tide is tiny compared to what would happen if a small object, like a comet, came to within a certain distance of a planet. In the nineteenth century, a French mathematician named Eduard Roche theorized that if a body came close enough to a planet, tidal forces would become so strong that the body would fall apart. On July 7, 1992, Comet Shoemaker–Levy 9 passed within Jupiter's Roche limit and was catastrophically disrupted.

COLLISION WITH JUPITER

Around April 10, 1993, the comet yielded another surprise. Now that observers had accurately measured the comet's motion over a period of time, Marsden published a circular announcing that the comet was in an orbit about Jupiter. Although astronomers had suggested that a few earlier comets might have orbited Jupiter for brief periods in the past, this was the first time that a comet had been observed to do this. Jupiter actually had 21 new moons.

On May 22, 1993, the world of astronomy was stunned by a new announcement: Shoemaker–Levy 9 would collide with Jupiter during the summer of 1994. For the first time ever, humans would get to see what actually happens when a comet strikes a planet.

TENSION BUILDS AS IMPACT APPROACHES

For the 14 months after the world learned of the impending impacts, astronomers worked feverishly to prepare every major telescope on Earth and in orbit about the Earth to observe this rare event and to predict what might happen. As impact day approached, that particular debate intensified: "The Great Fizzle is coming!" said one scientist. The comet would simply fall to pieces before impacting Jupiter, and Earthbound observers would see nothing.

A WEEK TO REMEMBER

Before July 16, much of what we knew about comet impacts was theoretical. On that day comet impacts entered the world of reality when the first fragment of Shoemaker–Levy 9, a fragment that was called A, forced its way into Jupiter's upper atmosphere at the astonishing speed of 135,000 miles an hour. A bright fireball surged upward some 2,000 miles above Jupiter's cloud tops—an enormous height, equivalent to the distance across the state of New York.

On Monday, July 18, a tiny fragment approached Jupiter. Its small flash was seen by at least one telescope. Some 30 seconds later, the gates of hell appeared to open as fragment G blew up over Jupiter. Its fireball soared more than 2,000 miles above the clouds.

The biggest surprise was yet to come, for Shoemaker–Levy 9 was not just a comet for the professional astronomers to study—it was for everyone to enjoy. By the evening of July 18, the G impact site was so dark that virtually anyone could see it through almost any telescope. Larger than the Earth, the impact sites of the G, H, K, and L fragments were clearly visible even to children looking through small telescopes. These were the most obvious features seen on Jupiter since Galileo first turned his telescope toward the giant planet in 1610.

When the last fragment hit on Friday, July 22, a large group of exhausted and ecstatic astronomers began the long task of interpreting what had actually happened. Because the amount of data was so large, and the kind of event we saw was so different, this will take years. The first task was to try to bring together the observations made with so many different telescopes, and then figure out what we actually saw with each impact.

EARLY RESULTS

What We Think We Know

Despite the fact that more observations were conducted of this astronomical event than of any other since people first looked up at the sky, at this early juncture there is actually little of which we are certain. An impact of this nature had never been seen, and the event so far has raised as many questions as it has answered. As months and years passed after the impact, some of these issues became more clearly resolved, giving way to clearer answers as well as further questions. In the questions that follow, we will indicate how those answers have changed over time.

How would you describe Shoemaker–Levy 9 to a nonscientist? A substantial majority of the scientists questioned said that they would describe Shoemaker–Levy 9 as a comet, not as an asteroid. Now, the weight of evidence is that the object was a comet whose orbit had been influenced by Jupiter for a very long time. The comet became a satellite of Jupiter around 1929.

What were we looking at when we saw the first flashes? Most scientists believe we were actually looking at the light from meteors deep in Jupiter's stratosphere that was being reflected off infalling dust.

Where did Shoemaker–Levy 9 originate? A majority of scientists suspect that the comet began its wanderings in the outer solar system, rather than near Jupiter or in the main asteroid belt.

What were the dark clouds made of? A slim majority thought that they were comprised of hydrocarbons, a sootlike material. However, a significant minority thought they were made up of silicate material.

The spectra of several metals were revealed during the impacts. Where did they come from? Almost everyone thought that they came from the comet, and not from Jupiter. On the other hand, more than half of the scientists thought that the sulfur was from Jupiter. That is an interesting find, for sulfur is thought to be the agent responsible for giving Jupiter its lovely colors—blue for lower features, red for higher ones. Also, sulfur has not been seen on Jupiter before.

What We Don't Know

How large were the largest fragments? Astronomers are still as curious about this as children are. Despite all the observations, we are still uncertain of how large the fragments were. Both before and after the impacts, theorists modelled the collisions using powerful computers. In one model, a comet only 600 meters across, hurtling into Jupiter at 60 km a second, could produce all the effects that were observed, including the enormous plumes. Other theorists disagree. One group suggests that 3 km across is the best size for the largest comet fragment. A third group suggests 2 km. One thing seems certain: if the comets were on the small size, then the Earth is in more danger than we previously thought from collisions from small comets. However, it might well be that these dramatic effects were produced by large comets. In that case, the Earth need worry about a catastrophic collision, such as the one that spelled the end of the reign of the dinosaurs, once every 100 million years on average.

Planetary scientists were similarly divided on how massive the largest explosions were—the assumptions were 10^{27}, 10^{28}, or 10^{29} ergs, a range from 10,000 to over 2 million megatons.

How were the fragments put together? Coherent, solid masses; fluffy snow; or loose, unconsolidated gravel? Scientists were almost equally split among these three possibilities.

How deep did the comets penetrate? Suggested answers were all over the map on this question, although new evidence has emerged that the largest ones might have gone down more than 300 km below the cloud tops.

What is the frequency of events of this magnitude on Jupiter? Although most impact experts thought that we were seeing a once-in-a-thousand-year event, the fact that this comet had the double performance of splitting up in an earlier encounter with Jupiter makes this type of event even more rare. We may never know how privileged we were to see so many comets cascade into Jupiter in just a few days, but the fact that dark spots of this number and intensity have never been seen on Jupiter since the invention of the telescope is good evidence that this was an incredibly rare event.

COMMONLY ASKED QUESTIONS

Of the many questions we have heard, these are the most frequently asked by children:

Did the comet make Jupiter explode or change its orbit? No. If you get stung by a mosquito while you are riding a bicycle, you will not fall off your bike, or even change the direction in which you are riding. Jupiter's plight was not unlike an attack by a string of mosquitoes. Even though they left some very noticeable marks, the comets were nowhere near massive enough to do any permanent damage to Jupiter or change its orbit in any noticeable way.

How will the events on Jupiter affect the Earth? At the time of the collisions, Jupiter was 477 million miles away. Although a lot of energy was released, none of it could possibly travel through space to reach the Earth. However, the same cannot be said about the enthusiasm that the Jupiter impacts generated here on Earth. They captured the imagination of many people, and it is hoped that a real result of the impacts on a planet far away will be a surge of interest in science on this planet.

Did the events on Jupiter tell us something about how the dinosaurs died? Possibly. As already mentioned, scientists have suggested that high temperatures followed the impact 65 million years ago, and that in the months after that impact a cloud of dust covered the Earth. Very high temperatures were recorded around the Jovian impact sites, and clouds much larger than the size of the Earth have persisted on Jupiter for several months.

How big were the spots? The largest spots exceeded 30,000 km in diameter within a few hours of their formation, which is much larger than the diameter of the Earth, and the spots were expanding at the fantastic rate of 4 km a second. No one predicted that the spots would be so big and so dark that they could be seen by virtually any telescope.

How long did the spots last? Although they had been spreading out, joining together, and fading, the line of spots was still a prominent Jovian feature as of the end of October 1994, when Jupiter finally got too close to the Sun to be observed from Earth. They were still visible as a single bar near Jupiter's south pole for a year after the impacts. Scientists also observed higher than expected levels of the gases hydrogen cyanide and carbon sulfide almost 2 years after the event.

43. *Nadine Barlow demonstrates how to make craters.*

In witnessing the collision of Shoemaker–Levy 9, we also watched the birth of a whole new phase of the science of impacts. For one special week, nature let us peer into one of her most closely guarded secrets. Comet crashes and life have gone hand in hand in the long history of the solar system, and that includes Earth, from the early times as life was gaining a foothold to the extinction of the dinosaurs 65 million years ago. In its destruction on Jupiter, Comet Shoemaker–Levy 9 gave us a precious reminder of our past.

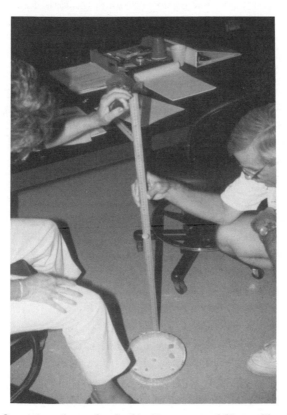

44. *Cratering shown by Jackie Unangst and Victor Sheaffer.*

DEMONSTRATION: SEEING CRATERS

The Earth has been hit often in its long history, and each hit resulted in a large crater. Over the passage of time most of the craters have disappeared owing to erosion and other geologic processes. But on the Earth's neighbor world, the Moon, the craters are still there. Encourage the children to look at the Moon through binoculars or a small telescope and count the circular features. Even the largest features, the ones that form the eyes of the Man in the Moon and can be seen clearly without binoculars, are the results of ancient impacts by very large objects. One of them, the Imbrium Basin, was carved out in a few minutes almost 3.9 billion years ago. The other eye, the Serenitatis Basin, is slightly older. Both basins have been filled with dark lava, which is what makes them so easily visible.

Through a small telescope, near full Moon you can see a large crater on the Moon's south side, surrounded by a system of bright rays. (Through most astronomical telescopes, south appears at the top of the reversed field of view.) Tycho is the name of this crater. From analysis of samples taken from a ray that crosses the Apollo 17 landing site, Tycho is thought to have been formed about 100 million years ago. It is about half the size of the crater that was blasted out at the end of the dinosaur reign.

CLOSURE: Discuss why the Moon is full of craters while the Earth has just a few, and Jupiter, being composed of gases, has none. After looking at hundreds of craters on the Moon, could we conclude that impacts have played a big part in the history of the Earth as well?

DEMONSTRATION: MAKING CRATERS

Using a cat litter box or an aluminum roasting pan, some gravel or sand, and a stone, we can show how craters form. Drop the stone into the box and watch the crater appear. Try tossing the

stone in from high and low angles and try different sizes of stones to see what differences these variables make in the crater's shape.

There is a more sophisticated way to make craters. Fill the box with two layers of different colored material or with a layer of sand topped by a layer of gravel. The material from the lower layer should fly out from the impact site and land again, in the form of rays such as those we see around Tycho. Also try using a slingshot to force the stone into the box at high speed.

Precaution: Emphasize to the children that this is a scientific experiment, and not a game. The slingshot is for making craters in the ground and must not be used against other children! No human has ever been hurt by a falling comet; let's keep it that way in the classroom. The other children should be kept out of harm's way until the stone hits its intended mark.

CLOSURE:　If a tiny stone has an observed effect on a sandbox, can you imagine how much effect an object from space about 60 yards wide, traveling at far greater speed, will have on the Earth? The resulting crater would be close to the size of Meteor Crater in northern Arizona. What about a comet more than a mile wide, coming in at a far higher velocity, hitting Jupiter, or hitting the Earth? We found out what happened on Jupiter in July of 1994, and the dinosaurs found out what happened on Earth some 65 million years ago.

Part IV

Beyond the Solar System

The purpose of Part IV is to introduce the distant parts of the sky in a simple way, including suggestions for sharing appropriate information with students.

This section is more informational in tone than previous parts of the book. Teaching the abstract and unfamiliar concepts of stellar and galactic astronomy to young students is much more difficult and often less appropriate than teaching the more concrete and familiar aspects of Earth-based astronomy and the solar system. While it is valuable for even young students to learn the patterns of the constellations in order to enhance their observing experiences, it is difficult for these same learners to grasp the concept of a sphere of gases millions of miles in diameter at a distance of hundreds of light-years from Earth.

Chapter 13

Stars Are People, Too

Moving beyond the Moon, beyond the farthest planets, and out into space, we finally arrive at the realm of the distant stars. Distant is the understated term here. Where we live in space is sometimes hard to grasp, and it is easier when we see our Earth as one of a group of worlds orbiting the Sun. But what about the multitude of stars in the sky, an array that seems to go on without end, stars that seem plastered on the velvet backdrop of the night sky?

Ancient cultures thought of the stars as part of a fixed sphere, and they joined them in various ways to make constellations that had meaning for them. The seven bright stars that circle the pole, for example, were a dipper to some, an Egyptian priest's magical hook, a plough, or part of a bear to others.

A colossal change in our understanding of the stars came in 1572, when the Danish astronomer Tycho Brahe noted a bright new star in the W-shaped constellation of Cassiopeia. Tycho couldn't believe his eyes. He was of a time that believed that the stars were unchanging points of light. What was he to make of the new star? With each passing night it grew brighter, until it was as brilliant as Venus and visible in broad daylight; it remained that way for weeks.

The star seen by those of Tycho's time was the supernova of 1572, an event that marked the end of a star's life. Tycho's star probably began its life as the smaller, blue member of a pair of suns we call a double star. It orbited about a much larger, red giant star. The small star's gravity was strong enough to pull a continuous stream of hydrogen from the large star, a stream that formed a disk of material around the smaller one. For millions of years the disk grew in size and in mass. As temperatures inside the disk rose, the star continued to gather material, creating a condition that was impossible to maintain, which would finally seal its fate.

We know of many double star systems that have a similar transfer of hydrogen from a large, cool star to a small, hot companion. But most of these stars blow off their excess hydrogen in giant thermonuclear explosions that take place from once every few months to once every few decades, and often once in several thousand years. Tycho's star had no such safety valve. For many millions of years the hydrogen kept building up, and when the explosion finally came, it destroyed the star completely. In less than a few weeks, the star brightened until it outshone the combined light of our entire galaxy, with its 200 billion stars.

Within a few months the supernova faded, and today there seems to be no mark in the sky where once a dazzling star ruled it. However, it left an indelible mark on history, for its light arrived on Earth just as the great civilizations of Europe were arising from centuries of intellectual sleep, a period we know as the Renaissance after the Dark Ages. People were beginning to look at their environment in new ways and were asking questions about the Universe. The supernova's light reached Earth only 29 years after Copernicus had proposed his view that the Sun, not the Earth, lies at the center of the Universe.

THE SUN'S PLACE AMONG THE STARS

With the exception of the Moon and a few bright planets, every point of light we see in the sky is a sun. Our Sun appears round and bright only because we are close to it. Each of these other suns would appear bright to inhabitants of their worlds. But there the resemblance ends. Virtually half of these stars are multiple, with two or more stars locked in a gravitational embrace. Some of these double stars are real unicorns in the stellar zoo—they are not even round, but teardrop-shaped, as their close-by companions pull their material away.

Thankfully our Sun is a very stable star. As gravity pulls the Sun's material toward its center, thermal pressure drives it outward, and so the Sun is said to be in equilibrium. Although it is larger than most stars, the Sun is about average in temperature and brightness, and it shines through a process called nuclear fusion. At its core, some 4 million tons of hydrogen are fused into helium every second. Although this has been going on for some 5 billion years, less than 6 percent of the Sun's supply of hydrogen has been converted into helium.

The Sun will retain this equilibrium as long as there is hydrogen in its core to provide fuel for its nuclear fires. Other stars at different stages in their lives show us what happens when the hydrogen in their cores runs out.

MAKING SENSE OF THE SPECTRA OF STARS

Early this century, two astronomers—Ejnar Hertzsprung from Holland and Henry Norris Russell from the United States—independently came up with a diagram that helps us to understand the diversity of stars by plotting spectral type against how bright, or luminous, a star really is, regardless of how far away it is from us. Most of the stars, including the Sun, plotted nicely on a section of the diagram called the Main Sequence. The red giants fell in the upper right-hand corner, and white dwarfs wound up in the lower right.

Most of the stars we see in the sky are called main sequence stars. They include the Sun, as well as stars like Altair, which are much brighter.

Red Giants

What happens the moment a main sequence star such as our Sun fuses its last gasp of hydrogen into helium? The star first has trouble when hydrogen in its core runs low. The core then begins to get smaller and the shell around it starts to burn its hydrogen faster. The more the core contracts, the hotter it gets, and the outer layers of the shell continue to expand. Over millions of years, the star gets brighter, larger, and redder. In a very brief process called the helium flash, helium fuses to carbon. No longer a main sequence star, it is now called a red giant. Aldebaran, the brightest star in Taurus the bull, and Antares of Scorpius, for example, are red giants.

During this process the star can start to vary in brightness. Some stars are redder than others: Hind's Crimson star (also called R Leporis), a star south of Orion in the constellation of Lepus, is so red that some observers liken it to a drop of blood in the sky.

White Dwarfs

As a star shrinks from its giant stage, it becomes a white dwarf, a very hot, compact star representing a late stage in the evolution of a star in the Sun's size range. But by the time a star has reached the white dwarf stage it has lost at least half its original mass. As these stars shrink they slowly turn color from red to white, and they are no longer fusing anything.

But these stars still shine! What keeps them going? It's a process called electron degeneracy. As these stars continue to get smaller from their red giant stage, their electrons resist the force of gravity, which is trying to push them even closer together. At a certain point the two forces balance each other out. The stars stop getting smaller, and over the next several billion years they survive as white dwarfs.

Inside a white dwarf, where a star the mass of the Sun is compressed into something the size of the Earth, the pressures are incredible. A cubic inch of white dwarf material would weigh some 15 tons here on Earth. But white dwarfs are very stable, and will shine peacefully for many billions of years. As their heat finally radiates away, astronomers suspect that they will stop shining altogether, becoming black dwarf stars. But even in the lifetime of the Universe, that process takes a very long time. If any stars have reached that stage, we haven't found them.

Red Dwarfs

In the Hertzsprung–Russell diagram, the main sequence stars fit nicely on a band that stretches from the upper left to the lower right. This makes sense, since most of the hot, blue stars shine more brightly than the cooler, red stars. All the main sequence stars are called dwarfs, not because they are necessarily small but because they are not giants.

At the red (right) end are the red dwarfs, the most common kind of star. If we could see all the red dwarfs the sky would be thick with them, and the graph would be heavy with stars on its lower right. Cool and small, these stars are stable for an incredibly long time. But the red dwarfs are intrinsically so faint that we can observe only the ones closest to us. Proxima Centauri, the nearest star to us, is a red dwarf.

The Supergiants

Very massive stars—with as much as 15 times the mass of the Sun—are called supergiant stars. These stars burn up their hydrogen at a much, much faster rate than do stars like the Sun. Close to 600 million miles across, Betelgeuse, the bright red star in Orion's shoulder, is an example. Rigel, also in Orion, is a blue supergiant only one-tenth the size of Betelgeuse. Instead of a 10-billion-year stable lifetime, these stars may use up their hydrogen in only 10 million years. But then they begin converting helium into carbon, and unlike ordinary stars, where there is only a helium flash, in the supergiants helium continues to burn after it ignites, eventually replacing the helium core with one of carbon. If the star is massive enough, the carbon core may itself start to contract, and fusion continues to produce cores of increasingly heavier elements until finally the star has a core of iron. As we shall see, stars in which this process goes further blow themselves up as supernovae.

THE LIFE CYCLE OF A STAR

This chapter has the title "Stars Are People, Too," because just as people have life cycles, so do stars. The birth of a star is a long process that begins with a vast cloud of mostly molecular hydrogen. It is very cold, its molecules of hydrogen hovering near absolute zero. Somewhere nearby, a massive star blows itself apart. The supernova spreads carbon into nearby space and into our giant molecular cloud. Gradually a fragment of this cloud starts to compress, and also to rotate about its center. Owing to gravity, its particles start to clump together, or accrete, very slowly as individual particles that settled quickly to the cloud's central plane find one another. At the center of the accretion disk is a body that will grow and grow until one day it ignites and begins nuclear fusion. All this took place when our Sun was "born" some 4.5 billion years ago.

A star such as our Sun has a very long stable period—maybe 10 billion years—as its nuclear fires consume its hydrogen supply. More massive stars consume their hydrogen faster. But in the case of any star, there is a last perfect day that ends as the star fuses its last atoms of hydrogen. (Our Sun, as we noted earlier, has at least 5 *billion* years to go before that happens.) Then the star turns its

45. *Teachers demonstrate the life cycle of a star.*

46. *Life cycle of a star. Photo courtesy of Marla Hensley, Kindergarten. Cavett Elementary School.*

process of nuclear fusion to its helium supply. As its helium starts to turn to heavier elements such as carbon, the star begins to swell into a red giant, like the star Mira. Then it will shrink. In the case of our Sun, for a brief time it will even come back to the size it is now, with what is left of the planets still circling it. As the star continues to shrink it becomes a white dwarf star about the size of Earth, its material so dense that a handful would weigh several tons. For billions of years it will shine as a white dwarf, until finally its light dims and goes out forever.

More massive stars have violent endings. After their nuclear fires form carbon, they continue to work, fashioning heavier and heavier elements. Finally the process results in the generation of iron. Iron will not fuse, and the process stops. The star's outer layers begin to collapse, and the star caves in on itself. Then with a final, tremendous release of energy comes the spectacular supernova explosion. The star brightens and brightens until it might outshine the combined light of its entire galaxy, and it can maintain that brightness for many weeks. The star's core is all that remains from this momentous event, a neutron star spinning very rapidly. If we can see it pulse each rotation, we call it a pulsar.

SUPERNOVA! THE 1987 STORY

Early in February 1987, astronomers were still resigned to the fact that there hadn't been a really bright supernova in our own galaxy since 1604. Astronomers have seen these exploding suns in other galaxies—several hundred of them in the last half century—but not one near home.

All that changed in 1987, when Ian Shelton, a Canadian astronomer at the University of Toronto's telescope in Las Campanas, Chile, began a photographic patrol for variable stars in the two galaxies nearest to the Milky Way, the Large and Small Magellanic Clouds. Hoping to discover the small stellar outbursts we call novae, Shelton noticed on his photographic plate an extra star near the Tarantula Nebula, a large region in the Large Magellanic

Cloud. The star was so bright that at first Shelton didn't believe it was real, but when he went outside and looked up at the cloud, he saw the new bright star.

At a neighboring telescope at the same observatory, Oscar Duhalde, an observatory night assistant, was checking the sky conditions when he too discovered the intruder. A few hours later, half way around the world in Nelson, New Zealand, a veteran amateur variable star observer named Albert Jones also detected the star, which was by this time shining brightly enough to be seen without a telescope.

Over the next few months the supernova shone on, brightening up until it was almost as easy to see as the North star, a hemisphere away.

Of all the phenomena astronomers could study with this rare event, one of the most exciting was the detection of the neutrinos that traveled from the site of the explosion. If these subatomic particles could talk, what a story they would tell! Their long journey actually started when they were released as the star's core was collapsing seconds before the explosion. But for a short time what was left of the core was so dense that the neutrinos couldn't go anywhere. Then the star exploded, and when the shock wave hit the neutrinos, it sent them cascading into space. Traveling at the speed of light, some of them reached Earth 160,000 years later.

BLACK HOLES

Being right in the middle of the Milky Way, the constellation of Cygnus, the swan, has lots of faint stars that we see only as the Milky Way's band of hazy light. One of these stars has the uncelebrated name of HDE 226868. It is a blue supergiant that is some 15 times more massive that our Sun. This star first came to astronomers' attention when they discovered that something at that position was rotating very quickly, once every few thousandths of a second. Astronomers using radio telescopes found a radio source they called Cygnus X-1. Turning their telescopes in its direction, all they found was the supergiant, bright enough to be seen in a small telescope. But no lumbering blue supergiant can rotate anywhere close to once in several milliseconds. Only pulsars—the rotating neutron stars that are the remains of a stellar core after a star has exploded as a supernova—rotate that fast. Could the neutron star be orbiting the blue star? It was when they calculated the orbit that they realized just how big this object was. Although the objects are orbiting each other, the invisible one is more than a quarter—and possibly as much as half—as massive as the bright blue supergiant!

A neutron star that massive would never stay a neutron star. It would continue to collapse indefinitely until it became as small as a city, a house, a doorknob, and then disappear altogether, leaving only a source of gravity so strong that even light cannot escape from it. Most astronomers conclude that such a thing—a black hole—is what resides next to HDE 226868.

DOUBLE STARS

Being a single star, our Sun is actually in a minority. More than half of the stars are multiple systems of two, three, or more stars orbiting each other. Many of these are delightful to look at through a telescope. Albireo, the second brightest star in the constellation of Cygnus, is a beautiful double star involving a bright yellow star and a fainter blue one. Their colors are sharp even through a small telescope.

The middle star in the handle of the Big Dipper, named Mizar, appears to be a bright double star. A faint star visible without a telescope on clear nights is nearby. These two stars are not related. The fainter one is much farther away and just appears close by because it is in the same line of sight. We call such coincidental double stars optical doubles. However, the brighter star is a real double star. Look at it closely through a telescope. It may appear elongated or even as two distinct stars. The two stars are almost equal in brightness.

THE MAGNITUDE SCALE

Our system of magnitudes dates back to Hipparchus, the second century B.C. Greek astronomer who divided the stars into six brightness groups. The 20 brightest stars were assigned first magnitude and the faintest stars sixth magnitude. By 1856 Norman Pogson of the Radcliffe Observatory quantified these classes; a first magnitude star is 100 times brighter than the faintest star visible without a telescope on a clear, moonless country night; a second magnitude star 2.5 times fainter than a first magnitude star; a third magnitude star, in turn, 2.5 times fainter again.

Vega, the brightest star in the Summer Triangle, is a zero magnitude star. Pogson defined Polaris, the North star, as being second magnitude. Most of the stars in the Big Dipper are also about second magnitude, and most of the stars in nearby Cassiopeia are a magnitude fainter.

Since the stars differ in their *apparent magnitude* for two reasons—intrinsic brightness and distance—wouldn't it help if we saw the sky with all the stars at the same distance? Imagining all the stars at the distance of 33 light-years, astronomers have defined *absolute magnitude* as the magnitude a star would have if it were that distance from us. Thus, any star that is 33 light-years away would have its absolute magnitude the same as its apparent magnitude. At this distance, our Sun would be a fifth magnitude star.

Comparing apparent and absolute magnitudes is a little like looking at lampposts on a street at night. Although all the street lights are the same strength, the ones closer to you appear brighter than the ones down the road. However, the lights in the shopping center parking lot at the very end of the street appear much brighter, even though they are farther away, because they are intrinsically more powerful.

To see how powerful each light is, imagine that all the lights, both on the street and at the shopping center, are at exactly the same distance from you, say halfway down the street. Then we could measure their absolute brightnesses, unaffected by distance.

The light that was outside your window now appears much fainter, and the distant shopping center lights appear brighter.

STAR COLORS

As a careful look skyward will show, the stars are not all white. Look at Antares in Scorpius, Aldebaran in Taurus, or Betelgeuse in Orion and you should see that these stars are reddish—not red like traffic lights, but a subtle red. Vega in Lyra and Rigel in Orion have delicate bluish shades.

A star's color is a clue to its nature. A blue color signifies that the star is much hotter than our Sun; red means cooler.

THE NEBULAE

"My God! There's a hole in the sky!" More than two centuries ago, the man who discovered the planet Uranus found a totally new kind of object. Herschel's idea was that these things might be "old regions that had sustained greater ravages of time" than their surrounding stars. They were not. Herschel and his son John, who imagined that these objects could be doorways to the infinity of space beyond, had found matter in space that hides the light from stars behind it. In the mid-nineteenth century the younger Herschel was cataloguing the southern stars from his observing site to prepare for his General Catalogue of Nebulae. Deep in the southern sky he found an egg-shaped dark region that looks so black we now call it the Coal Sack.

One of the first things we notice about a dark, country sky is that the stars in it are not uniformly distributed. Even the Milky Way, where the stars are so thick we see them as a band of light, has an irregular shape to it. We can even see bands of darkness—bands which are really large clouds of mostly hydrogen gas. Unlit by nearby stars, the clouds are invisible except when their presence blocks out the light from more distant suns. The most obvious one straddles the Milky Way in the Northern Hemisphere constellation of Cygnus.

Spread throughout the galaxy, these dark clouds provide the raw material for new stars. If we look toward the sword of Orion, we can see where new stars are actually being formed out of the surrounding gas. The gas is a part of the galaxy in which we live, and because there is so much of it blocking the light from more distant suns, it was only recently that we could figure out where in the galaxy our home actually is.

Thus Herschel's holes in the sky are not portals to the great beyond but simply nebulae. But because there are no nearby stars to light them, they are dark. Edward Emerson Barnard, the American comet discoverer who was one of the most prolific observers ever, catalogued a large series of dark nebulae. He found them in two types of places in the sky; in the Milky Way, where we see them because they block out the light of more distant stars, and together with bright nebulae, where absorbing dust blocks the light of the brightly lit gas behind it.

Unless a dark nebula lies near a bright one, we have a great deal of difficulty figuring out how far away it is. If a nebula blocks out part of the stars from the more distant Milky Way, we know that it is closer than those stars.

Because they actually glow, the emission nebulae are the most colorful representatives of these clouds. Through a large telescope (say 10-inch-diameter or more) and under a good sky, some show beautiful tinges of red and green. Sometimes the clouds consist mostly of dust, which shines simply by reflecting light from the nearby stars. Merope, one of the stars in the Pleiades, is surrounded by a good example of a reflection nebula. You cannot see it without a telescope, but on a clear, dark night out in the country, a small telescope should reveal this cloud of the Pleiades.

BOK GLOBULES

Almost 50 years ago, the famous astronomer Bart Bok disclosed that the Milky Way was filled with dense concentrations of gas that appear to be the birthplaces of new stars. These tiny globules captured the public imagination when a 1957 science fiction story by astronomer Fred Hoyle had a globule swallowing the Earth. Decades later, the HST photographed similar objects. Called EGGs (for *e*vaporating *g*aseous *g*lobules) these pillar-shaped objects are being irradiated by ultraviolet light from nearby stars. The energy from this ultraviolet radiation is causing the globule's gas to evaporate.

STAR CLUSTERS

More than a hundred mighty globular star clusters, each containing tens of thousands of stars, lie scattered throughout the sky. Through a small telescope they look like small fuzzy spots, but larger ones resolve the spots into a large number of stars. Globular clusters have been studied almost since the invention of the telescope—in 1665 Abraham Ihle found a large cluster, now called Messier 22, in Sagittarius. By 1786 William Herschel suggested that these clusters were large groupings of stars.

Most of the globular clusters we see are in the Milky Way's outlying regions. The Southern Hemisphere has a monopoly on the best of these distant giant hives of old stars. Omega Centauri, a large oval clustering of hundreds of thousands of stars, is prominent there, although it can be seen on late spring evenings from the southernmost parts of the United States. But 47 Tucanae, the finest globular of all, lies too far south to be seen from the United States. Messier 13 in Hercules, the Northern Hemisphere's best globular, is 23,000 light-years away. There are globular clusters in the galactic plane, but we don't see the distant ones because intervening dust clouds block off their light. A globular cluster can be 100 light-years wide.

Among the oldest things in the galaxy, the stars of the globular clusters are almost as old as the galaxy itself. Some estimates put them as old as 16 billion years.

On the other end of the age scale is another type of cluster called the open or the galactic cluster. (We call them "open" because we can see their individual stars more clearly than we can see the globulars, and galactic because most of them are near the plane of the Milky Way.) Since the ones we see are much closer than the globulars, we see them not as fuzzy spots but as masses of individual stars. If you've seen the Pleiades, you've seen an open cluster. The Pleiades are also known as the seven sisters; why is a mystery, since there are only six members that are bright enough to be seen with the naked eye. Maybe a seventh has faded since ancient times. But through a telescope you can see many fainter stars.

We are passing through an open cluster right now. It is a loose association of stars called the Ursa Major moving cluster. (We call it moving because the rates and directions of motion of its stars have been measured.) Most of the Big Dipper stars are part of this array. We know that so many far-flung stars belong to a single cluster because we can measure what we call their proper motions, which give us an idea of the speed and direction that the stars are moving. So when astronomers found that a large number of stars in our vicinity of space were moving with a common speed and direction, we concluded that these were part of a moving cluster.

OUR PLACE IN THE GALAXY

Our galaxy contains some *200 billion* suns, many of which we never get to see because there is also so much dust and gas that obscures our view. Generally, the galaxy is as flat as a pancake, except for the area around the center, where there is a wide bulge.

Surrounding the center, the galaxy looks like a pinwheel, as several spiral arms uncoil to a distance of over 100,000 light-years. But that's not all—surrounding this galactic disk is a halo that stretches at least the same distance further out. The halo contains the globular clusters, some of which we can see through binoculars from our own backyards. Stretching even further out, and including the space some small neighbor galaxies occupy, is a thin layer called the corona.

SPIRAL STRUCTURE

What does our galaxy look like from the outside? Trying to describe its spiral shape has daunted scientists for years. It's like trying to paint a picture of the outside of your house when all you've ever seen is a room or two from the inside. But by observing the nearby houses from your window, maybe you could get a clue as to what the outside of your house looks like.

We now know that our galaxy is a pinwheel-shaped cauldron of stars. During World War II, Walter Baade used the 100-inch telescope at Mount Wilson, California, to discover that the stars in the Andromeda Galaxy were of two types, one that predominated in the galaxy's center and the other in the arms. He also identified regions of hydrogen gas that were common in the spiral arms but not in the center. Our galaxy has similar populations of stars.

In 1950 William Morgan of Yerkes Observatory observed stars of these two populations to discover that the Milky Way has at least three spiral arms, one going off into Orion, another in Perseus, and a third in Sagittarius. Since then studies by radio astronomers have traced the galaxy's arms much further out.

CONCLUSION

Whether we enjoy the mythology of the constellations or study the astrophysics of the stars that make them up, the stars of our galaxy offer much to grab our attention. Get the children into them slowly, at their own pace, and they'll remember what you tried to teach them and enjoy it always.

ACTIVITY: MAGPIES AND THE MILKY WAY

BACKGROUND: The story of the "Weaving Princess and the Herdsman" (or shepherd or farmer) can be found in both Chinese and Japanese traditions. Vega, the star represented by the weaving princess, is by far the brightest star in the constellation Lyra, the harp. Altair, the star represented by the herdsman, is the brightest star in Aquila, the eagle. Vega and Altair are part of an asterism, or group of stars, called the Summer Triangle. The third star in the Triangle is Deneb, located in Cygnus, the swan, or the Northern Cross. Another famous asterism is the Big Dipper, with its seven prominent stars within the constellation Ursa Major, the Great Bear.

OBJECTIVES: The students will learn about an Asian legend and festival, will practice an Asian art form, and will learn about three constellations, an asterism, and the Milky Way.

PROCEDURE:

1. Share the story about the Milky Way and the magpies with the class.
2. Mount the figures of the weaving princess and herdsman on oak tag or thin poster board, cut them out, fold and affix the stands with tape. Option: Decorate the figures with crayons or markers.
3. The science portion of the lesson can center on constellations and asterisms; these stars are easily visible in the fall.

STORY SUMMARY:

Long ago there was a beautiful princess, the daughter of the Sky God. She was the most skillful weaver in the land, weaving beautiful cloth at her loom every day. One day she looked up from her loom and saw a herdsman at work. She fell in love with him at once. When the herdsman saw the princess at the window, he also fell in love.

The weaving princess begged her father to allow her to marry the poor herdsman. The Sky God agreed, and the two were very happy together. They were so happy that they each neglected their work. The princess forgot to weave her beautiful cloth, and the herdsman neglected his animals. The Sky God decided to punish them.

The Sky God placed the princess in the sky in one place and the herdsman in the sky in another place. Between them he put a river of stars. They could see each other, but could not cross the river. The princess and herdsman returned to their work with great sadness.

The Sky God took pity on them and decided that if they worked hard at their tasks, he would allow the princess and the herdsman to meet one night each year. Toward the end of summer, a great flock of magpies flew to the river of stars. They settled onto the water and formed a bridge for the princess and herdsman to cross. The next night the magpies were gone, and the princess and herdsman returned to their work for another year.

CLOSURE: If you are fortunate enough to live near a dark site, encourage the students to observe the Milky Way and the Summer Triangle and its constellations. Share stories about the Milky Way from other cultures.

Chapter 14

Advanced Project: Variable Stars

Of all the secrets that the stars hold, perhaps the most interesting is that some of them, called variable stars, change in brightness. There is even an organization called the American Association of Variable Star Observers (AAVSO), in existence for more than 80 years, that has as its sole purpose the collection of observations of these variable stars by amateur astronomers.

A variable star, as the famous amateur astronomer Leslie Peltier once noted, is not just a star that's there, it's a star that's happening. Take Mira, a red star in Cetus, the whale. Whatever brightness it has when you first start to watch it, it won't keep for long. Although the star might be quite bright when you first meet it, with each passing week it will fade until you cannot see it without binoculars, and then you will need a telescope to watch it dim to its minimum of ninth magnitude. For a week or two Mira will just hover at this low point of its cycle, and then it will slowly begin to recover, brightening through the telescope, then through binoculars, and at last it will shine again as a star fully visible to the naked eye. Eventually it will reach third magnitude or brighter.

Mira's period of variation, from one maximum to the next, takes about 11 months, and is called a long-period variable star. Mira represents a phase our own Sun might go through when it gets older—some 5 billion years in the future—and finally runs out of hydrogen. It will swell in size, eventually swallowing up what is left of Mercury, Venus, and the Earth. (If you tell this story to children, keep reminding them that these drastic things will not happen for a very long time, and that they should not worry about the Sun's losing its reliable ability to light and heat our world.)

There are other types of variable stars. In the constellation of Perseus, Algol, the "Demon Star," normally shines at second magnitude, the second brightest star in the constellation of Perseus. For a few hours every two days, the star fades by a magnitude as a fainter companion star passes in front of it. Then as the fainter star moves past, Algol brightens again. Algol, the most famous example of an eclipsing binary, is interesting partly because it is so famous; its variability, easily followed with the naked eye, has been known for many centuries. It is bright, and if you know when the next convenient minimum will occur, you can plan to observe it. Since Algol's drops in brightness do not often happen during evening hours, a careful teacher can schedule an observing party when there is to be an evening fading. These events are listed in magazines such as *Sky and Telescope* and in the annual *Observer's Handbook of the Royal Astronomical Society of Canada*.

Delta Cephei, a star in Cepheus, the Ethiopian king, changes in brightness over a period of 5 days. It is not an eclipsing binary system; rather this star varies because of a change that takes place within it: As the star expands it fades, and as it contracts it brightens. Since the star remains bright throughout its cycle, it can be followed each night with the naked eye or a pair of binoculars.

Betelgeuse, the bright red star in the northeast corner of Orion, is a good star to observe. It is bright and red and part of a major constellation, so finding it is easy. Its strange name also might attract a child's attention. However, its pattern of change is very slight and very slow.

SOME HISTORY

Understanding the motivations of people who devoted their lives to observing variable stars is as important, and as inspiring, as watching the stars themselves. The life of the British observer John Goodricke is especially interesting: Unable to hear or speak, he let the night sky become an important part of his life. He measured the period of Algol and discovered the variations of Delta Cephei. Tragically, shortly after he discovered Delta Cephei's interesting behavior, he caught a bad case of pneumonia and died at the age of 21. Through his story, the variable stars and those who study them can come closer to children.

Like people, stars behave in many different ways. In teaching children about the sky, variable stars—stars that change in brightness—are a key to understanding a sky that performs. Young children can enjoy observing variable stars if the stars perform regularly and are easy to find. Most important, variable stars teach us the lesson of hope, that the starry sky is a different place from the earthly environment of a child, and that it is accessible to young people. By following the antics of a variable, we share in the secrets of the sky above us.

ACTIVITY: OBSERVING A VARIABLE STAR

AGE: This project is intended for middle school students or older.

PURPOSE: To experience the changes in light from a variable star.

MATERIALS:

A pair of binoculars
Star chart: Figure 47
Pencil and paper

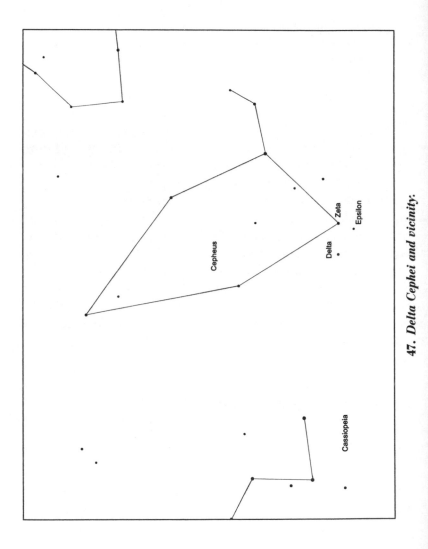

Cassiopeia

Cepheus

Delta

Zeta

Epsilon

47. Delta Cephei and vicinity:

PROCEDURE:

1. Find where Cepheus is in the northern sky. Since it is circumpolar from most northern latitudes, it should be easy to find, unless it is low in the sky.
2. Find Delta at the end of an isosceles triangle with two other stars, Epsilon and Zeta.
3. Using binoculars, look at the three stars in the triangle. Delta is the star at the apex. Look now at the other two.
4. Which is the brighter of the two stars? We will assign the number 1 to the brightness of that star. To the fainter one we assign the brightness value of 5.
5. We will be estimating the brightness of Delta using a 5-point scale, from 1 to 5, where 1 equals the brightness of Zeta and 5 equals the brightness of Epsilon.
6. Look carefully first at Epsilon, then at Zeta, then at Delta.
7. Is Delta nearly as bright as Zeta?
8. Is Delta nearly as faint as Epsilon?
9. Now make your brightness estimate:
   a. If Delta is as bright as Zeta, give it a 1.
   b. If Delta is as faint as Epsilon, give it a 5.
   c. If Delta appears exactly midway to brightness between Epsilon and Zeta, give it 3.
   d. If Delta is just a little fainter than Zeta, give it a 2.
   e. If Delta is just a little brighter than Epsilon, give it a 4.
10. Record the observing session and the result in your observing logs.
11. Congratulations! You have just made your first estimate of the brightness of a variable star. It is part of a great tradition begun in the middle of the nineteenth century; the AAVSO has collected several million observations since it was founded in 1911. If you like, you can even translate your estimate to the traditional system of magnitudes. Since Zeta Cephei is 3.6, and Epsilon Cephei is 4.2, you can interpolate roughly as follows:

1 = 3.6	2 = 3.8	3 = 4.0	4 = 4.1	5 = 4.2

However, this extra step is not necessary to appreciate the idea of the brightness change that is taking place in this star.

12. Repeat the entire procedure each night for 2 weeks.
13. Try plotting a graph showing the changing brightness of Delta Cephei with time. Put the numbers 1 through 5 along the y-axis and dates on the x-axis.

CLOSURE: For the first time, your group has tried a research project in astronomy. Was it fun? What did we learn? What if we take on another type of variable star next time, one that is less predictable? What if a star erupts like the supernova Tycho saw? Could we use the same procedure to estimate the brightness of the newly discovered star?

Observing variable stars can be a lifelong activity. By following the brightness changes of these stars, by participating in science, you are getting a taste of what astronomical research is really like.

OTHER TYPES OF VARIABLE STARS

Besides eclipsing, Cepheid and long-period variable-behavior stars can vary in other ways. Every few weeks, a star called SS Cygni, usually at twelfth magnitude, erupts and increases in brightness by more than three magnitudes in a few hours. SS Cygni is an example of a *dwarf nova*. It consists of a large cool star orbiting a small hot star, and a stream of hydrogen gas that travels to the small star and forms a disk around it. After a critical amount of hydrogen surrounds the small star, it explodes, and the star brightens. T Coronae Borealis, a recurrent nova, erupts every few decades. Not far from T is R Coronae Borealis, a nova in reverse: Usually visible through binoculars at a bright sixth magnitude, at irregular intervals it plunges to a faint fourteenth magnitude and cannot be seen except in moderate-sized telescopes. Apparently this star occasionally releases a giant sootlike cloud that surrounds it and blocks its light.

As a variable star observer, it would be your pleasant task to follow the progress of a selection of variable stars. Algol's minima can be found in the *Observer's Handbook of the Royal Astronomical Society of Canada* (available from the Society at 124 Merton Street, Toronto, Ontario M4S 2Z2, Canada).

You might even try out a membership in the AAVSO. They send each new member a beginner's kit consisting of charts for ten stars. Their address is:

American Association of Variable Star Observers
25 Birch Street
Cambridge, Massachusetts 02138

Chapter 15

Galaxies and the Universe

At the beginning of chapter 13, we moved out beyond the Moon, beyond the planets, and beyond the distant sphere of comets until we reached the realm of the stars. Now we resume our outward journey. Imagine traveling beyond the stars, the nebulae, remnants of ancient supernovae, and the outskirts of our own galaxy to the void that lies beyond. Here is the realm of the galaxies, where the sky is marked not by the pinpoints of light we call stars but by giant islands in the sea of space, each a cauldron of billions of stars, each called a galaxy.

Throughout this book we have emphasized how, as centuries passed, our perception of the Universe in which we live has changed. Copernicus suggested that Earth was not the center of the solar system, and Galileo, we learned in chapter 9, was able to prove it. In the 1920s the American astronomer Harlow Shapley demonstrated that we do not even live in the center of our galaxy, but out in its boondocks, many thousands of light-years away from the center. Also in the 1920s, Edwin Hubble showed that our Universe of galaxies is expanding.

It is easy to see a single galaxy—ours—just by looking up. We live in the midst of a grouping of stars called the Milky Way, which includes every star we can see in the sky with the unaided eye. When he pointed his telescope at the heavens, Galileo saw individual stars in the Milky Way. We are not at its center but at its edge. All the stars we see throughout the sky are relatively close to us. The farther stars in our arm and the neighboring arms of the galaxy are bunched together in the Milky Way. However, when we look in the direction of the constellations of Scorpius and Sagittarius, we are looking into the center of our galaxy. Much of the galactic center, as well as the spiral arms that lie beyond it, is obscured by large amounts of dust. Even though the Andromeda Galaxy is made up of hundreds of billions of stars, it was not until the 1920s that telescopes became powerful enough to detect them.

It also happens that the Milky Way is far more active a galaxy than was previously thought. At its center might lie a supermassive black hole with the mass of a thousand suns and untold amounts of energy, out of which even light cannot escape. We live in a spiral galaxy, one of the larger of the hundreds of thousands that have been seen across the Universe. Our home galaxy contains some 200 billion suns, all in different stages of development. The last two times we know of that stars exploded in our galaxy were in 1572 and 1604 (one occurred in the Large Magellanic Cloud, a neighboring galaxy, in 1987), but there is evidence for another explosion that took place in Cassiopeia. Though no record exists of a visual sighting of this event, the remains have been found as a radio source called Cassiopeia A. Smaller stellar explosions are seen within our galaxy several times each year, and there are small eruptions every few months in hundreds of stars.

In the autumn Northern Hemisphere sky, we can see the brightest of the external galaxies. Lying near the stars Beta and Mu of the constellation of Andromeda, this galaxy is known as Messier 31, or the Andromeda Galaxy. Older textbooks, which indicate that the Andromeda Galaxy is twice as big as the Milky Way are mistaken; it turns out that our own galaxy is bigger than was previously thought, being about the same size as the Andromeda Galaxy. We are part of a small cluster of galaxies called the "Local Group," which contains some 15 galaxies. Although most of the member galaxies are small, the group contains three great spiral galaxies: the Milky Way, Messier 31 in Andromeda, and Messier 33 in Triangulum.

Moving even farther out, we find that our Local Group belongs to a throng of groupings of galaxies, which we call the "Local Supercluster." Many of the member galaxies can be found in the constellations of Ursa Major, Coma Berenices, Leo, and Virgo. This area contains, in fact, more than 13,000 galaxies! If you point a small telescope into the center of the Virgo cluster and then move it about slowly, you should see several galaxies move through your telescope's field of view.

With so many galaxies around, it would seem that they would be of many different kinds, but actually there are only a few varieties. The giant *spiral* systems are a common type, and large telescopes have found them by the thousands. Another type is *elliptical*: Messier 87, containing perhaps 5 trillion stars, is one of these. Younger galaxies are often *irregular* in type. In Ursa Major there is Messier 82, a galaxy we call *peculiar*. Divided across its width by several large dark bands, it seems that this galaxy is undergoing a giant burst of new star formation.

HOW THE GALAXIES ARE DISTRIBUTED IN SPACE

A look skyward will show that the stars in our galaxy, or at least in our neighborhood of our galaxy, are not distributed evenly. When Edwin Hubble of Caltech did his early surveys of the galaxies in the 1920s, he concluded that they were spread pretty evenly in space. However, during his long search for distant planets in the 1930s, Clyde Tombaugh of the Lowell Observatory discovered that the galaxies were not distributed evenly at all, but were clumped together in gigantic groupings. A decade later, George Abell of the Palomar Observatory confirmed this idea, and went on to show that the clusters themselves are clumped into vast superclusters. The number of stars in all these galaxies is virtually without end; it is even said that the number of stars in the Universe is greater than the number of grains of sand on all the beaches of Earth.

A REALLY STRANGE UNIVERSE!

As we move even farther out into space, we find galaxies stretching out without end. The Universe seems to go on and on, and distance seems to get harder and harder to imagine. The farthest distance we can actually visualize is probably that between the Earth and the Moon: Many children know of cars that have 100,000 miles on their odometers; that is about two-fifths of the Earth-Moon distance. The farthest thing that we can see in the night sky is the Andromeda Galaxy, whose distance has been pegged at 2 million light-years. It is impossible to imagine a number that large. Light travels 186,272 miles every second; that is the equivalent of about seven times around the world each second. At that speed, light takes 2 million years to reach us from the Andromeda Galaxy; in other words, the light reaching us from Andromeda today left there when the earliest humans were walking about the Earth.

As far as that galaxy is, it is one of the closest. Using the Hubble Space Telescope, astronomers have sighted galaxies as far off as 10 billion light-years. In fact, one of the *quasars*, which is probably the very bright, active nucleus of a distant galaxy, is bright enough, at magnitude 12.7, to be visible through small telescopes. That means that it is possible to look through a small telescope and see out to the very edge of the known Universe.

The quasar story became famous in 1963, when the moon passed in front of a quasar, a radio source called 3C-273. Astronomers do not normally know the precise location of sources of radio energy in the sky, but the occultation allowed astronomer Cyril Hazard to calculate its position exactly. Then, using the 200-inch reflector on Palomar, Maarten Schmidt found a fairly bright object at that location. He photographed the object's spectrum, and the result stunned him: It seemed that the object was at an extreme distance and rushing away from us at a small fraction of the speed of light.

Things get ever stranger as we go farther out. In 1979, three astronomers, D. Walsh, R. Carswell, and R. Weymann, used the University of Arizona's multiple mirror telescope to find evidence of a *gravitational lens* in space. The first hint came with the appearance of two identical quasars, one right next to the other. To have real quasars joined up like that by coincidence would be virtually impossible, astronomers thought, so they concluded that perhaps they were seeing a single quasar that was very far away. As the quasar's light made its way here, it passed a big galaxy. The galaxy's gravitational field was so strong that the quasar's light was split and bent its way around the galaxy, so the quasar appeared as double from here. A more magnified view that revealed a faint galaxy in between the two images of the much more distant quasar proved the theory. Nature has since shown itself capable of providing even more interesting visual imagery: a quasar's light split into four images. Called Einstein's cross, this beautiful, esoteric feature in the sky appears as four images of a quasar surrounding a central galaxy. Even more recently, HST images have shown giant arcs of light. To find them, the telescope was turned toward densely packed groups of galaxies to see if more distant galaxies would have their light perfectly bent around the galaxies, thus appearing as arcs.

HOW THE UNIVERSE BEGAN

Although there is some disagreement, most astronomers accept Hubble's law, which says that the Universe is expanding and that it began as a single explosion. Published in 1929 and dubbed the "Big Bang" by Sir Fred Hoyle, a famous English astronomer who actually intended to ridicule it, the idea suggests that the entire Universe—matter and energy—was compressed into a single point. In a huge explosion the Universe began to grow, an expansion that continues to this very day. The great superclusters of galaxies began to evolve quite soon after the Big Bang, and over time the Universe took on the appearance it has now.

ACTIVITY: MAKING AN EXPANDING UNIVERSE

PURPOSE: To give the children an idea of how the Universe is expanding.

MATERIALS: A balloon
Thin-tipped marking pen
A set of powerful lungs!

PREPARATION:
1. Inflate the balloon.
2. Mark dots at random on the balloon.
3. Deflate the balloon.

CARRYING OUT THE EXPERIMENT:
4. Note the crowding of dots on the surface of the uninflated balloon.
5. Inflate the balloon.
6. Observe how the dots appear to rush away from each other as the balloon inflates.

CLOSURE.   If you imagine that the dots on the balloon are galaxies in space, the expanding surface of the balloon is making the galaxies appear to rush away from one another. However, the surface of the balloon is only a two-dimensional model of an event that is three-dimensional. A more realistic way to visualize the expanding Universe in a classroom (but one that is harder to accomplish) is to bake a loaf of raisin bread. As the loaf rises, the raisins, representing the superclusters of galaxies, appear to move away from each other. (The individual galaxies are not rushing away from each other, nor are the clusters of galaxies. It is rather the superclusters of clusters of galaxies that are moving away from each other.) As an example, from the point of view of one raisin, as the bread rises another raisin 1 inch away will move to 2 inches away, while a raisin 2 inches away will move to 4 inches away. The farther raisin moves twice as far in the same time, that is, it moves twice as fast.

Comparing dots on a balloon or raisins in bread to the monstrous galaxies that make up our expanding Universe is an exercise that is somehow unfulfilling. The truth is that the Universe is an unimaginably big place, and the events that occur, as well as the way we see those events, are incredibly complex. What we can do using the balloon or the raisin bread analogies is to give the children a sense of the process, if not the vastness, of the Universe of which we are a part.

OBSERVING THE BRIGHTEST OF THE DISTANT OBJECTS

The distant objects we have just written about are not just esoteric things that no small telescope user can ever see. Scattered throughout the sky are 110 "deep sky objects," which were collected into a catalogue by a famous eighteenth-century French observer Charles Messier. Some of them are nearby galaxies and others are clusters of stars or clouds of gas and dust.

Messier wanted his claim to fame to be the comets he discovered. The first person to find comets as part of an organized telescopic search, Messier learned how to find comets on his own. From 1760 to 1770 he was virtually the only discoverer around, so prolific that Louis XV of France called him the Comet Ferret. He was granted a steady income that let him search for comets to his heart's delight.

Messier's good fortune ended with the storming of the Bastille and the start of the French Revolution in 1789. Although there were still a few comets left for him to find, he was without the pension that allowed him to live while he spent his time hunting for comets. Virtually penniless, he even had to borrow oil for his observing lamp from one of his friends.

During all these years Messier kept a listing of cometlike objects that he found during the course of his search. The list included objects within our own galaxy—clouds of gas and dust and clusters of stars—as well as other, distant galaxies. Today, Messier's comets are long gone, but the entries in his catalogue are forever present in the sky. Not all of them are external galaxies, but they are all distant objects.

ACTIVITY: FIND THE MESSIER OBJECTS

PURPOSE: To gain experience learning to use a telescope and to learn about the different types of distant objects the sky offers—for children in high school or advanced middle school.

MATERIALS:
A telescope equipped with a finderscope
A chart showing the Messier objects and their location

PROCEDURE:
This is a long-duration project designed to last the entire year. Each season, on clear, moonless nights, the student should try to locate several new Messier objects.

1. Using the chart, find the location of the Messier object.
2. Align the telescope's finder with the main telescope by looking first at a bright star or planet through the main telescope and then adjusting the finder's screws so that the object is also centered in the finder.
3. Use the telescope's finder to "star-hop" from star to star, comparing what you see with stars on the chart, until you arrive at the position of the Messier object. It should appear in (or near) the field of the main telescope.
4. Take notes on the appearance of the object. Draw it if possible.

CLOSURE: By the end of this project you will have sighted a number of the objects that Charles Messier recorded 200 years ago.

Your knowledge of the deep sky will have improved vastly, and you should be quite comfortable using a telescope.

THE MESSIERS: A LIST OF THE BEST

Objects marked with a * are part of our galaxy. Messiers marked with the symbol ** are technically part of our galaxy, but are located at its edge. Other galaxies are marked with a ***.

Northern Hemisphere Spring

**Messier 3:   A glorious globular cluster of stars, very nice even through a small telescope. About midway between the stars Arcturus and Cor Caroli (heart of Charles), this object is not hard to find.

*Messier 11:   The Wild Duck cluster of stars. Glorious cluster, with one bright star near its richest part. However the bright star is not physically a member of the cluster.

**Messier 13:   The Great Cluster in Hercules. The finest globular cluster in the northern sky. When you look at this cluster, you are seeing its 100,000 stars as they appeared 23,000 years ago. M13 is that many light-years away.

*Messier 44:   The Beehive Cluster in Cancer is one of the prettiest clusters of stars in the entire sky. Faintly visible to the naked eye on a good night, this cluster contains some bright stars that appear to be arranged, coincidentally, in triangles.

***Messier 51:   The Whirlpool Galaxy. Found near the end star of the Big Dipper's handle, this is a fine example of a galaxy seen face-on. It looks double because a second galaxy appears to hang on from one of the arms

***Messier 64:** The famous Black-Eye Galaxy. Through moderate-sized telescopes on a good night, you can see a large dust area that looks like a black eye.

***Messier 65 and ***Messier 66:** Two spiral galaxies close together, near the tail end of Leo.

***Messier 83:** A beautiful galaxy in Virgo, near Corvus. It has a stunning appearance if seen through a large telescope under a dark sky.

***Messier 87:** One of the largest galaxies in the heavens, M87 is an elliptical galaxy containing perhaps 5 trillion stars.

***Messier 88:** Richly beautiful galaxy in Virgo.

***Messier 100:** Spiral galaxy in Virgo. In 1979 a bright supernova appeared in this galaxy.

***Messier 101:** Spiral galaxy in Ursa Major.

***Messier 104:** The Sombrero galaxy in Virgo, a beautiful galaxy that actually bears a resemblance to a Mexican hat.

Northern Hemisphere Summer

Note that few galaxies are visible during the summer because we are looking at the plane of our own Milky Way, so our view of other galaxies is blocked.

Messier 4: A beautiful, loosely structured globular star cluster in Scorpius.

Messier 5: Large globular cluster in Serpens.

*Messier 6:** Large open cluster in Scorpius.

*MESSIER 7:   Large open cluster in Scorpius.

*MESSIER 8:   The Lagoon Nebula in Sagittarius. A rich region of gas, part brightly lit, part dark, which includes many young stars.

*MESSIER 16:   The Eagle Nebula. A rich nebula, filled with stars and clouds. In 1995 the HST revealed the presence of EGGs in M16.

*MESSIER 17:   The Omega Nebula. Perhaps the finest nebula in the sky, Omega looks like a colored checkmark.

*MESSIER 20:   The Trifid Nebula. An amorphous blob in small telescopes; in larger ones dark lanes stretch across its length, dividing it into three areas.

**MESSIER 22:   Globular cluster in Sagittarius. One of the largest in the Milky Way.

*MESSIER 27:   The Dumbbell Nebula. This is the cloud that surrounds a star that shed its outer layers long ago. It is called a planetary nebula.

*MESSIER 57:   The Ring Nebula. This planetary nebula is one of the easiest Messiers to find in the entire sky. It is midway between the two bottom stars of Lyra.

Northern Hemisphere Autumn

*MESSIER 1:   The Crab Nebula. Messier's first object turned out to be the interesting remains of the supernova that exploded in 1054.

**MESSIER 2:   Aquarius Cluster. A beautiful globular star cluster. Appears as a bright spot of light through a small telescope, but larger telescopes reveal the mottled appearance that indicates that it is a cluster of stars.

**MESSIER 15:   Pegasus Cluster. Another gem of a globular cluster, one of the best in the autumn sky.

***MESSIER 31:   The Great Galaxy in Andromeda. By far the most beautiful galaxy in the sky. On a clear dark night away from city lights M31 can be seen with the naked eye; it is in fact the farthest object usually visible without a telescope.

*MESSIER 45:   The Pleiades. A fine open cluster, clearly visible to the naked eye.

Northern Hemisphere Winter

*MESSIER 35:   Open cluster in Gemini. Rich with stars.

*MESSIER 42:   The Great Nebula in Orion. One of the sky's most exquisite regions, this nebula is easily visible as a misty spot in the center of Orion's sword.

**MESSIER 79:   Although this globular cluster is faint, we include it because it is the only accessible globular in the winter sky. It is located south of Orion in Lepus. Globular clusters tend to lie close to the plane of our Milky Way, which is most prominent in the summer sky, so they are not common in the winter sky.

***MESSIER 81:   Large, striking spiral galaxy in Ursa Major.

***MESSIER 82:   Close to M81, this unusual galaxy has several dusty areas blocking our view. Astronomers believe that the galaxy is undergoing a burst of star formation.

Two examples from Charles Messier's catalogue give us an idea of its richness. The cluster Messier 13 is 23,000 light-years away at the fringe of the Milky Way, but Messier 31, the Andromeda Galaxy, shines at us from a distance of 2 million light-years. By showing us distant objects in our own galaxy as well as far-off galaxies, this catalogue gives users of small telescopes a sense of the majesty of the Universe. For young people with perseverance and a sense of adventure, there can be no better project than the observation of some of the objects on this very old list.

Chapter 16

Searching for Life Out There

Are we alone in the Universe? What if there is intelligent life on other planets? Among the most common questions we have heard from children, especially older children, these, which address the issue of life elsewhere, reflect a central theme and deserve careful answers. While the discussion that follows is a little more complex than the general level of this book, we have found that children are so interested in the subject that they absorb the material quite well.

We know for sure only that there is one star in one galaxy that has one planet that supports intelligent life. That star is our Sun, in its place in the back alleys of the Milky Way, which nurtures the life we see around us on Earth. How would we go about looking for other examples of life? What do we look for? What do we listen for?

FINDING LIFE ZONES AROUND STARS

In our solar system, the Earth is 93,000,000 miles from the Sun, and the temperature here is just about right. Venus is too close to the Sun and too hot, and Mars too far and—at least at present—too cold. So around the Sun, we have a narrow life zone where a life-supporting planet can orbit. That zone would be closer to the star than we are to the Sun for cooler stars and farther for hotter or larger stars.

The notion of life zones, however, considers only stars that are basically right for life. Our Sun is actually not the ideal star— its heavy dose of ultraviolet radiation would make life impossible here were it not for a narrow ozone layer high in our atmosphere that protects us from the radiation. If our Sun were a bit redder, for example a G5 or even a K star, then life would actually be more comfortable here, and we wouldn't need to worry about sunburn.

THE SCALE OF THE UNIVERSE

We do not know if we are alone. But if we are, then we must live in an extremely large and mostly empty house. Let's see just how far out of touch we are just from the center of our own galaxy, let alone other galaxies that stretch out across the Universe.

Most of the stars we see each night are relatively close to us, on the order of several hundred light-years away. At the speed of light, 186,000 miles a second (as we have noted before, that is the equivalent of swinging around the Earth seven times every second) these distances seem vast. But how vast are they really? Take the Summer Triangle, the popular asterism of Vega, Deneb, and Altair. Vega and Altair are bright because they are close. Altair is only 16 light-years away, and Vega 26. Although it seems to be almost as bright as the others, Deneb is actually a far more luminous star, a gigantic sun that shines from the vast distance of 1,600 light-years.

Our galaxy is a big place. To travel from one end to the other at the speed of light would take close to 100,000 years. The two nearest galaxies, the Magellanic Clouds and the Andromeda Galaxy, are, respectively, about 160,000 and 2 million light-years away from us, and more remote galaxies are billions of light-years away. Of all the stars in all these galaxies, most are incapable of supporting life as we imagine it. The red giants and white dwarfs we have already considered are not candidates for harboring living worlds.

WHAT ARE WE LOOKING FOR?

When we think about life on other worlds, we must first define what sort of life we mean. Any life at all? That might be too broad a search. It is possible that there are life-forms so different from ours that we would not recognize them as life even if we were to bump into them at a supermarket. So we qualify that by searching for life "as we know it"—an idea that involves a search for something that might relate to any of the millions of species already inhabiting Earth. We can narrow this further and restrict our search to "life we can communicate with." Even that presents a problem. We know that humpback whales and dolphins are life-forms on Earth with large brains and complex patterns of communication. We share our planet with them, but still have no idea how to communicate with them other than on the most basic level. If a spaceship loaded with Denebian dolphins were to land here, we could not share their thoughts, hopes, or plans, nor could we learn much about these aquatic travelers.

There is a further point. The earliest life-forms that appeared on Earth were simple one-celled creatures resembling bacteria. Over time, successive generations of creatures came along, each more complex than its predecessor.

Even if a planet had the perfect conditions that could lead to the start of a simple life-form, how would that life-form evolve? Because evolution is a random process, chances are it would not have any closer resemblance to us than we have to bacteria. They might be complex as we are, but almost certainly infinitely different.

These days the world of science fiction is filled with aliens that have a head, two eyes, two ears, and a mouth that speaks English. The popular science fiction show *Star Trek* justified this by invoking a fictitious law of related development, which states that similar planets would produce similar cultures, but in the real Universe there is no evidence that that would happen.

Notice that we have been narrowing down the parameters of our search for extraterrestrial life. We can actually quantify this narrowing activity and perhaps come up with an estimate for the number of planets in our galaxy that could support life as we know it and life we can communicate with. Radio astronomer Frank Drake offered a way to do this with an equation so simple that even a sixth-grader can follow it. However, we on Earth will probably never know its answer.

The equation is:

$$N = N_* f_p N_e f_l f_i f_c f_L$$

The magic number N will be the number of civilizations within our galaxy with the means of contacting us. We know that N equals at least 1: the Earth. But can N be more than 1? On the other side of the equal sign, we start with the number of stars in our galaxy, N_*, which is about 200 billion. But only a small fraction of these stars will have systems of planets surrounding them. Let's represent this fraction, a number between 0 and 1, by f_p. Now it is possible to say that if just one in 200 has a planetary system, there are a billion such systems.

Three such possible systems have been detected since 1984. In that year, University of Arizona scientist Brad Smith used satellite information and images from a telescope to show that the southern star Beta Pictoris has a disk-shaped cloud around it that could be a planetary system in formation. A few years later, scientists detected a system of three planets around a pulsar. Now, neither of these systems is anywhere close to supporting life as we know it—the Beta Pictoris system might resemble our solar system as it was some 5 billion years ago, and the three planets circling the pulsar have had violent lives, most likely even being involved in a supernova explosion that altered both the planets and their orbits drastically. In 1995, astronomers discovered a wobbling motion in the nearby star 51 Pegasi—a star similar to our Sun—that could indicate that a Jupiter-sized planet is there. However, this planet orbits so close to its sun—completing a year in only 4 days—that it could not possibly be in a life zone. Several other objects may also have been detected around 51 Pegasi, and it may turn out that some of these are better candidates for supporting life as we know it.

However, there may be some planets or moons with environments suitable for intelligent life. From our own planet's fragile ecology and the harsh environments of our solar system's other worlds, we infer that that number is quite low. We could say that each solar system might have only one planet with such an environment. Let us call that number N_e, where the e stands for ecologically sound for life. That doesn't mean that the planet is exactly like the Earth; it just means that on such a planet life could evolve. That gives us *a billion* possible planets. But life will not arise in all these environments. Let us say that it has arisen on only a tenth of the planets where it could have started. That fraction is f_l, the fraction on which life did get a foothold. We are now down to *100 million* possibilities.

This figure gives us the number for all life-forms. Only a very small fraction of all that life could evolve to at least our level of intelligence. We can call that fraction f_i for intelligence. We do not know what that fraction is, so let's guess a tenth once again. We are now down to *ten million*. Once such higher-order life is present, will it develop a technology to communicate with other intelligent beings? The humpback whale is very intelligent, as we have already noted, but we have no evidence that the species has any technology. Let's use the fraction f_c for communication in a technological form, such as radio. Again, we have no idea what that fraction is. Let's say a tenth again. We're down to *one million* possible planets.

The last consideration is time. For how long during the lifetime of a planet will such a beautiful civilization last? Civilization on Earth is several thousand years old, but we have had radio for only a century. So we need to take this question into account. We add f_L for the length of time. If that number turns out to be once again a tenth, then the number of civilizations we can look for is only *100,000*.

Are there really 100,000 civilizations in our own galaxy? That seems like a big number, but our galaxy is very large. The number could be wrong at any or all of the stages we've described. Some scientists specifically don't agree that f_i or f_c—intelligence and communication—should be as high as a tenth. If life appears, why should intelligence and technology follow? After all, of the virtually uncountable numbers of species on Earth now and in the past, only one seems to have developed that ability. The dawn of rational thought, and the ability to project it into space, might be an extremely rare and precious event. There could be but 10,000. Or 1,000. Or 100. Or just 10. Or 1.

THE EFFECT OF COMETARY COLLISIONS—AND OF JUPITER

It is quite possible that our Earth has other things going for it. The solar system happens to be crowded with asteroids and comets; we have already discussed the effect of collisions with these bodies. One point that we have not really considered is that after a major collision, the playing field is cleared and a burst of new species of life can occur. Had the Earth not been hit so often, complex life-forms such as ours might never have developed.

On the other hand, what would be the effect of too many collisions? When the Earth was young it was bombarded with comets and asteroids so often that life could not have gained a foothold. But Jupiter's enormous gravity acted as a giant vacuum cleaner, yanking all that cometary material away from the inner part of the solar system. In most cases the material was thrust out of the solar system permanently. Less often, comets would collide with Jupiter, as Comet Shoemaker–Levy 9 showed us so dramatically in 1994. Had Jupiter not been there, the Earth would be a target for a cometary collision far more often that it is, probably often enough to render life here untenable.

HAVE WE BEEN VISITED?

We can only guess at the real answer for the Drake equation. We can also wonder what would happen if somehow another civilization were to visit us. As we discussed earlier, in 1937 that possibility was the theme of a dramatic broadcast on the radio called *War of the Worlds*. In that program, which was directed by Orson Welles, unfriendly Martians landed and set about launching an attack. About 15 years later we were visited again, in the movie *The Day the Earth Stood Still*, by aliens who came in peace but we attacked them. By 1968, celestial visitors were coming in peace: *2001: A Space Odyssey* portrayed an intelligent civilization actually seeding life here.

What if there is a close encounter? The last thing we should expect is a parade of little green hominids on a main street in town. We should look for less obvious indications of an extraterrestrial visit. We might suddenly find a big new satellite in orbit around the Earth, a satellite not in any known log of artificial satellite launches.

What if a UFO—an *unidentified flying object*—lands in your backyard? Then what? Do you call 911, or NASA? Most scientists would be curious, but still skeptical. They would want to touch it, chip at it with a geologist's hammer—and when they finally opened the door, the occupants inside would have to be very different from two-eyed, single-headed hominids before we would finally admit that we are definitely not alone. It is important to emphasize that there are lots of rumored, but no confirmed, reports of any such objects landing anywhere on Earth.

LOOKING OUT THERE

Maybe we don't need to wait for an actual visit. Maybe we can look for evidence of artificial structures around nearby stars. The Earth gets only a tiny portion of the Sun's energy; most of it is lost in deep space. The physicist Freeman Dyson imagined a civilization so advanced that its people constructed huge spherical structures to contain all of a star's energy to provide power and life-support for a host of worlds. In a sense a Dyson sphere is an immense house, with a star at its power center.

Or we can look for messages from space. A message coming from another civilization, even a nearby one, might take hundreds of years to get here. A civilization at the far end of our own galaxy might have to wait tens of thousands of years before its message reaches us. Some astronomers have even looked at the nearby galaxy Messier 33 in the constellation of Triangulum to see if a galaxy-wide communication net had been set up there; none was found.

Even if another civilization were to send a message, what kind of message would it send? Whether we wanted to or not, we've been sending messages from Earth ever since Marconi sent the first radio waves. All kinds of radio and television shows have been broadcast into space. A planet 100 light-years away might just now be hearing our signals.

Two deliberate messages have already been sent to the stars. In the mid-1970s the enormous parabolic dish of the Arecibo radio telescope was opened in Puerto Rico. Built in a valley a kilometer wide, this dish can detect extremely faint signals. On its "opening night," however, it was used to transmit a signal, a simple message beamed toward M13, the globular cluster in the constellation of Hercules. That signal will take more than 20,000 years to get to the stars there.

Also in the mid-1970s, two spacecraft—Pioneer 10 and Pioneer 11—were launched toward Jupiter and Saturn. After its mission, each craft was to leave the solar system and head out toward the stars. Each was fitted with a beautiful brass plaque containing a sketch of what we look like, plus drawings showing where we live in our galaxy and in our solar system.

In 1977 Voyager 1 and Voyager 2 were launched toward the outer solar system. These craft are also heading toward the stars. Each Voyager has a simple digital record made of gold that contains music—from Chuck Berry's *Johnny B Goode* to the sounds of the humpback whale, as well as pictures of life on Earth. Who knows but that some other civilization will some day interpret the whale sounds.

It is unlikely that any of these messages will ever be read. The secret is that they were meant for the sender—they were intended for the people of Earth, in the hope that we will be able to learn that there is other life in this Universe besides our own, that we are part of a bigger picture that includes other stars and galaxies.

UNIDENTIFIED FLYING OBJECTS

During the decade of the 1950s there appeared to be a rush of sightings of what quickly became called *unidentified flying objects* or UFOs. Sightings were reported of flying objects that looked like saucers or cigars. Although these reports were common, very few came from people who were familiar with the night sky and the secrets it holds.

Of course, a UFO is simply something flying that has not been identified. A lot of these reports turned out to be misidentifications of bright planets, especially Venus when it shone brightly in the western evening sky. Two searchlights projected against high, thin clouds can make a good UFO, and they fooled David Levy one night in 1963. Strange effects such as ball lightning, or a military test flight, or a rocket launch from Vandenburg Air Force Base or White Sands in the western United States might be enough to convince someone that a visit is under way. But in almost every case UFOs can be dismissed pretty easily.

At the height of the UFO craze of the 1950s the U.S. Air Force opened Project Blue Book, an attempt to list and categorize all the UFO reports. Because the Air Force found, as did many astronomers, that there was simply nothing that could not be explained by natural phenomena, the project was eventually abandoned.

Maybe we should stop looking, and listen instead. That was the idea of radio astronomer Frank Drake, who in 1960 began a simple search for radio signals that might be repetitive, indicating an artificial origin. He listened in vain to several Sunlike stars, such as Epsilon Eridani and Tau Ceti.

Early in 1968 Jocelyn Bell Burnell, a graduate student in England, was watching incoming data from a radio telescope experiment when she noticed some signals that repeated with startling regularity. At first the thought of an artificial source for the signals was considered; they even called them LGMs for *little green men*. As it turned out, the real answer was just as astonishing as if we had been contacted. These signals were coming from the remains of a star that exploded in ages past as a supernova. They had found the rapidly rotating pulsar at the center of the explosion; with each rotation the star would pulse sharply.

By the 1970s, the search for extraterrestrial intelligence was an idea whose time had apparently come. We could, it was felt, even be part of a "galactic club" of civilizations, which might someday welcome us into membership. NASA even bestowed the acronym SETI, for Search for Extraterrestrial Intelligence, on its program. Very few nationally funded search projects ever got off the ground. One that did was due to the effort of Harvard University's Paul Horowitz, who made use of a beautiful radio telescope dish. The telescope had originally been built by astronomer Bart Bok in the mid-1950s to study the Milky Way. Funded by the Planetary Society, this search began at Harvard in 1983. A more ambitious program using Australia's Parkes radio telescope got under way some 10 years later.

WHY LOOK?

We want to look for other civilizations because it is our nature to be curious. Are we alone? Are we the only intelligent life-form in the Universe? Maybe. Maybe not. Our galaxy alone is so big, and civilizations possibly so fleeting, that we may never know another. Could we ever visit these other places? Will we someday discover the secret of "warp drive" that would allow us to cross the galaxy in years instead of millennia? They do that in *Star Trek*, but that is a TV show. Scotty is right, however, when he insists that "I canna change the laws of physics." We cannot travel faster than light, and so we cannot even in theory reach the very nearest star in fewer than 4 years of travel.

In 1895, Percival Lowell thought he had found the answer to the canals he thought he saw on the planet Mars. Mistakenly thinking that they were artificial structures whose purpose was to move scarce water across a dying planet, he hit upon the idea that a Martian civilization had created a global society with no racism, no conflict, and no war. Lowell was wrong about Mars. There has been no sign of any life there at all, let alone an advanced technological civilization. But he may not be wrong about the prospects of intelligent life, maybe on as many as 100,000 planets throughout our galaxy.

More important, Lowell was right about yearning to see other civilizations, life-forms better than ours, and cultures that can teach us. Our galaxy may be teeming with civilizations, but they are all so isolated from each other by huge distances that for practical purposes, each exists by itself.

As the sky darkens on a clear night, go outdoors with your children and look up at the hundreds of stars you can see. Possibly a life-form on a planet circling one of these stars might be looking back, wondering what we wonder, asking the same questions that we ask. We may never know this life-form. But what a special joy it is to think about this as we watch the sky from Earth, the one precious place where we know that life thrives.

Part V

Children Ask Questions

Chapter 17

Questions about Earth, the Sky, and the Solar System

Children are inquisitive, and when the subject is their environment, they have all kinds of questions. In this chapter we explore some that we have encountered, a sampling of the many that children will toss your way. In trying to answer them, the most important thing to remember is that all questions are good ones—there is no such thing as a stupid or silly question.

QUESTIONS ABOUT SKY PHENOMENA

Why is the sky blue?

This is a very commonly asked question. Sunlight that hits the Earth has components from all colors of the spectrum, but the atmosphere scatters blue light more effectively than longer (redder) wavelengths.

Is the Moon really bigger when it is near the horizon?

No, it is not. This effect is called the "Moon illusion," and it is thought to be caused by our seeing the Moon (or the Sun or a group of stars) relative to familiar objects on the ground. The Moon seems larger, but when it is high in the sky, it seems to be smaller because we have no landmarks to compare it to. It is easy to prove that the Moon does not change size. Look at it through a telescope when it is high in the sky, and again, using the same telescope and eyepiece, when it is rising and setting. It will be the same size. Some people have suggested to try looking at the Moon while bending over so that you are looking out between your legs. Your head is upside down, and you do not see the Moon illusion. Although this should work, it is so uncomfortable that we do not recommend it!

How can you tell if the Moon is getting bigger or smaller?

In its orbit around the Earth, the Moon does change its distance from the Earth, so that at some times the Moon does appear to be larger than at other times. However, you can't really notice this change without measuring it. The change is most apparent during eclipses of the Sun. If the Moon is near apogee (its farthest position from the Earth) it does not completely cover the Sun, so we see an annular eclipse, where a ring of sunlight surrounds the Moon. When the Moon is closer to perigee (its closest position to the Earth) it covers the Sun completely, and we see a total eclipse (see chapter 8).

How can you see meteors when they're so small?

As we explained in chapter 10, a meteor is an event that we see because a tiny particle is racing through Earth's upper atmosphere, causing the surrounding air to heat until it is incandescent. It is the reaction to the surrounding air that we see, not the tiny particle.

Why don't the stars shine in the daytime?

They do. However, our own Sun makes the daytime sky so bright by comparison that we cannot see any stars.

Why don't you see the same stars all night long?

Because the Earth rotates, or turns around on its axis, the stars appear to rise, move across the sky, and set. So do the Sun and the Moon.

Have you seen anything strange in the night sky?

One night in 1963 David Levy, then 14 years old, saw two bright lights race across the sky. Although he was puzzled at the time, he now thinks they were searchlights projected against clouds (see chapter 16).

QUESTIONS ABOUT BEING ON THE EARTH

How do you know Earth is round?

In chapter 9 we offer several lines of evidence to show that the Earth is a planet in space. To show that it is round, we need more specific ideas. (1) During an eclipse of the Moon, the shadow of the Earth is projected against the Moon. That shadow is clearly round, indicating that the Earth, its source, is round also. (2) The classic test, found in many textbooks, is to look for a sailing ship moving out to sea. If the Earth were flat, you would see it shrink and shrink until it became too small and disappeared. Because the Earth is curved, however, you see the ship appear to "sink" as it moves farther away. First the hull disappears, then the lower sails, and finally the upper mast vanishes.

That explanation is fine in theory, but it doesn't often work in practice. Refraction and other effects of the thick volume of air near the horizon cause the ship to assume all kinds of strange shapes that confuse this test. The ship might appear double, part of it might stretch, or it could assume some other grotesque shape. In any event, this is not a test we would recommend to children at the beach without the caveat that the Earth's atmosphere could play tricks with the experiment.

How come people on the bottom of Earth are not upside down? Why don't they fall off?

Because the Earth's gravity pulls us down, our sense of balance sees the ground as down and the sky as up. For us on Earth, what is up and what is down is a matter of perspective. If you are in an airplane that turns upside down for a moment, your mind rebels at the contradiction and you feel queasy or sick. Although it is hard to believe, there is no up and down in space. So wherever we are on the Earth, up is toward the sky and down is toward the center of the Earth. People in Australia are in no more danger of falling off the Earth than people in the United States.

QUESTIONS ABOUT TELESCOPES

Why does a telescope make things appear upside down?

The natural way for a telescope to work is to make objects appear reversed. If you want the telescope to show things right side up, you need to add an extra "erecting" lens or prism; binoculars work this way.

Which is better, a reflector or a refractor?

In chapter 5 we discussed the different types of telescopes, how they work, and which are more appropriate for a beginner. It is impossible to say which of the major types is better, especially since both lenses and mirrors—and combinations of the two—have had distinguished histories and have made many contributions to our understanding of the Universe. Years ago, inexpensive reflectors were far cheaper than refractors of the same size, but today there is less of a difference.

QUESTIONS ABOUT SPACE TRAVEL

How do you propel yourself in space?

Newton's third law of planetary motion says that for every action, there is an equal and opposite reaction. It is a consequence of that law that lets us travel through space or through the air on a jet plane. When we fire a rocket in one direction, we move in the opposite direction. In space, we could do anything to summon that third law and propel ourselves, including throwing something. If we threw a ball in one direction, we would go flying off in the other.

Is it possible to exceed the speed of light?

No, it is not. Warp drive, the science fiction idea of traveling faster than light, does not exist in any law of Nature that we understand. We cannot cross to the other end of the Milky Way in 70 years; at the speed of light it would take us 30,000 years just to reach the center of our galaxy.

Why can't you land a spacecraft on all the planets?

Only the four innermost planets, Mercury, Venus, Earth, and Mars, have solid, rocky surfaces. Pluto has an icy surface. The giant planets Jupiter, Saturn, Uranus, and Neptune are made of gases; they may not even have solid surfaces to support a spacecraft. When the Galileo probe entered Jupiter's atmosphere on December 7, 1995, it could not survive the temperature and pressure of the surrounding gases. After about an hour the probe disintegrated as it was expected to do. It did not find a solid surface (see chapter 9).

QUESTIONS ABOUT THE MOON

Why is the Moon upside down or on its side?

See chapter 8. This is a commonly asked question—asked in this way—but it is not really easy to answer. When the Moon is in its crescent or quarter phase, it appears to stand on its cusp when it is high in the sky. How it appears when rising or setting depends partly on its phase and partly on the latitude from which you are observing it. In the Northern Hemisphere, in temperate latitudes, the Moon sets at an angle. At southern latitudes, it appears to rest on its back when setting. The Moon's angle relative to the horizon is an effect that depends on where the Moon is relative to the observer's horizon.

Why does the Moon sometimes shine in the daytime?

The Moon actually spends half of its time in the sky during daylight hours. Only on the night of full Moon does it rise near sunset and set near sunrise. With every passing night the Moon rises anywhere from 20 minutes to more than 1 hour later, so naturally when the Sun rises the Moon still has a way to go before it sets, and therefore it appears in the daytime sky. At new phase, the Moon is in the sky only during daylight but we can't see it—except during an eclipse—because the Sun is shining on the far side and the side we see is not lit. Still rising later each night, as the Moon becomes a crescent it rises after sunrise, so that it appears in the afternoon sky and much more vividly in the western sky after sunset.

When the Moon is a crescent, why can you see the dark part?

This effect is called *earthshine*. When we see a crescent Moon, someone standing on the near side of the Moon would see an almost full Earth. Just as the Moon provides a faint glow to the nightside of the Earth, the Earth provides a brighter glow to the nightside of the Moon.

How is the Moon lit?

The Moon gets its light from the Sun. We see the Moon because it reflects sunlight.

Why does the Moon appear bright at some times during the day and then get dimmer?

When we see the Moon high in the sky during daylight hours, it appears bright if it is seen in a clear sky. As the Moon gets lower, its light travels through a denser layer of air, so it appears fainter.

QUESTIONS ABOUT THE SOLAR SYSTEM

How hot is the Sun?

The Sun's surface is about 11,000 degrees Fahrenheit. Its temperature rises to several millions of degrees at its center (see chapter 7).

How did the Sun get so hot?

The Sun formed as the center of a large cloud of hydrogen. As the cloud got denser, it grew hotter. For millions of years the temperatures rose because of the force of gravity at the center of the cloud. Finally the heat and pressure became so great that nuclear fusion began—the Sun ignited and became a star. After ignition, the Sun became even hotter and has been that way for some 5 billion years.

How many pennies would it take to fill the Sun?

This was a magical question from a second-grader, no doubt familiar with the problems of guessing how many pennies it would take to fill a jar. We presented the problem to Clyde Tombaugh, discoverer of Pluto, and it became a central topic of conversation at a weekly lunch of senior professors at New Mexico State University. The answer:

> 4.66 *decillion* pennies, or:
> 4.66 times 10 to the power of 30, or 4.66×10^{30}
> the total earnings of 4.66 *septillion* millionaires!
> If it were for sale it would cost 466 *novillion* dollars.

How were the planets made?

Some 5 billion years ago, the solar system was a large cloud. Nearby, a large star, finishing its life, exploded, sending carbon and other elements into the cloud. The cloud began to rotate. As the center grew heavy, the rest of the cloud organized itself into rings, which condensed into smaller balls of gas and dust. Once the center got hot enough, it ignited to become the Sun, and the smaller balls became the planets (see chapter 9).

How much would you weigh on the other planets and the Sun?

If you weigh 100 pounds on Earth, you would weigh on:

Mercury	38	Mars	39	Neptune	114
Venus	91	Jupiter	251	Pluto	7
Earth	100	Saturn	107	Sun	2795
Moon	17	Uranus	90		

How did the planets get round?

The force of gravity tends to pull things equally toward the center of an object. For objects larger than about 1000 miles across, the force of gravity is stronger than the material comprising the object, so the object tends to be round in shape. A smaller object with weaker gravity might have an irregular shape.

Are there clouds on every planet?

This question, asked by a perceptive second-grader, needs to be answered in two parts: first, which planets have atmospheres that would support clouds, and second, what these clouds are like. Although Mercury has no atmosphere to speak of, observers have occasionally reported clouds. Venus is totally enshrouded with clouds. Mars has white, yellowish, and even bluish clouds. Jupiter, Saturn, Uranus, and Neptune are totally covered with clouds of methane and ammonia. When Pluto is closer to the Sun, it warms so that some of its normally frozen ices turn to gas and become part of a temporary atmosphere.

Why is Venus as hot as Mercury?

When the tiny Mariner 2 spacecraft sped past Venus in 1962, its small package of instruments reported that the planet's surface temperature was about 800 degrees Fahrenheit. That spacecraft is still orbiting the Sun, its instruments silent forever, but its report left us with a vivid picture of an uninhabitable planet. Venus is the victim of a runaway greenhouse effect that leaves the planet hotter than Mercury, even though it is almost twice as far from the Sun. In a greenhouse, heat from the Sun warms up the room, but the glass prevents the heat from escaping. On Venus, heat from the Sun reaches the surface, but the large amount of carbon dioxide in the planet's atmosphere acts like the greenhouse glass, preventing heat from escaping.

It is important to understand the other planets so that we can understand the Earth. Venus teaches us what can happen when excessive amounts of carbon dioxide form in an atmosphere. Because of the materials produced by our own civilization, the amount of carbon dioxide in our own atmosphere is increasing. Although it should never get as out of control as it is on Venus, the greenhouse effect is very serious and should be studied carefully.

Why is Jupiter so colorful?

Jupiter's rich colors are believed to come from the presence of sulfur compounds high in its clouds. The sulfur causes low features to appear blue, and high ones, like the red spot, to appear red.

Why can't we stand on Jupiter?

Even though Jupiter is a gas planet, it is more than 10 times the diameter of Earth; if Jupiter were hollow, more than 1000 Earths would fit inside. Thus the volume of Jupiter's material gives it a gravity much greater than that of Earth.

Why are there rings around Saturn?

Saturn's rings are thought to be the result of a small moon of Saturn coming too close to Saturn and breaking up owing to the giant planet's gravity, or from the collision of a comet or asteroid with a moon of Saturn (see chapter 12 for a discussion on the Roche limit, and how it determines when an object will break apart).

What are Saturn's rings made of?

The beautiful rings are apparently made of tiny particles of icy material. Thin rings have also been discovered around Jupiter, Uranus, and Neptune (see chapter 9).

Can you stand on Saturn's rings?

No. There is probably a good deal of space between the particles that compose the rings.

What is a falling star?

See chapters 10 and 11. A falling star is not a star that has fallen out of the sky. After you see a meteor, there is not an empty space in the sky where a star used to be. A meteor is caused by a piece of dust, probably no larger than a grain of sand, that has entered the Earth's atmosphere. Friction with the air causes the grain to vaporize in a flash of light, and it is this light—not the tiny particle—that we see as a meteor.

What is the difference between a comet and an asteroid?

This question is rather difficult to answer, for comets and asteroids are related objects in our solar system. Basically, comets appear as fuzzy objects in the sky because they are surrounded by dust and gas. The word comet comes from the Greek word for long-haired. Asteroid means starlike, and asteroids do resemble starry points of light that move slowly through the sky. However, when a comet turns itself off and stops producing gas and dust, it appears as an asteroid. What we think is an asteroid can suddenly "turn on" and behave as a comet. A few times, an object that appeared to be an asteroid when it was discovered suddenly "turned on," as in the case of Chiron, which behaved more cometlike than asteroidlike as it came nearer to the Sun (see chapter 10).

When a meteorite hits, is it hot?

People who have picked up freshly fallen meteorites—within an hour or two of their fall—tell us that they are quite warm to the touch but not extremely hot. However, some meteorites have even had frost condense on them. As a meteorite passes through the atmosphere, its exterior rapidly melts away, almost as fast as heat can get conducted into its interior. Moreover, the object has been in the cold of space for millions of years, so its interior is extremely cold (see chapter 10).

Is it easy to tell if a rock is a meteorite?

It is not easy for someone without a lot of experience to tell whether a rock is a meteorite. Through careful observation, it is possible to tell the story of how a meteorite survived its flight through the atmosphere. Most meteorites have an uneven surface with a dark crust formed as their outer parts melted on their journey through the atmosphere and then cooled again after landing on Earth.

   For more advanced students: Some fallen meteorites called *siderites* are rich in iron or nickel and are reddish in color. But most red rocks are ordinary sandstone. Other meteorites, called *aerolites*, contain lots of diverse minerals such as silicon and magnesium oxides. In some aerolites, called *chondrites*, these metals are compressed into lumps we call chondrules. *Siderolites*, yet another meteorite type, contain both metal and stone.

Has anyone ever been hit by a meteorite?

Although heavy meteorites do not rain down on Earth every minute, they do hit often enough to cause some spectacular sights. On April 26, 1803, a shower of thousands of small stones fell over northwestern France, terrifying many people. On November 30, 1954, a fist-sized meteorite crashed through the roof of a house in Alabama, bounced off a radio, and injured the startled woman who lived there. Although there are no reports of anyone being killed by a meteorite, it is possible that a few people were killed by the great crash of June 30, 1908, over the Tunguska River in Siberia.

Finally, the town of Wethersfield, Connecticut, has seen two meteorite hits within about 10 years (see chapter 10).

QUESTIONS ABOUT ECLIPSES

Is it really dangerous to view a solar eclipse?

It is always risky to look at the Sun directly. When the Sun's light is diminished because the Moon is partially covering it, our eyes do not feel the need to squint; they remain fully open, allowing the full strength of the Sun's ultraviolet rays to enter and harm the retinas. Also, during an eclipse we *want* to view the Sun; our attention is drawn to the event so we tend to concentrate on it, making the damage much worse. Special eclipse glasses with filters, or minimum thickness No. 14 welder's glasses, add protection to your eyes (see chapter 7).

Is it dangerous to view a lunar eclipse?

No, it is not. An eclipse of the Moon is always safe to view (see chapter 8).

Chapter 18

Questions about the Universe, Life, and Philosophy

QUESTIONS ABOUT THE UNIVERSE

What is a black hole?

In chapter 13 we discussed the life cycle of a large star. As the star becomes a supernova, its core starts to collapse. Most stars will collapse to a certain point, then continue to exist as neutron stars or pulsars. But if the core was massive enough, it will continue to contract indefinitely until it reaches a point when its own light cannot escape from it. It is then called a black hole.

Why do stars have different colors?

In chapter 13, we looked at the different types of stars. Stars have different colors because they have different temperatures. Red stars are relatively cool, yellow ones such as our Sun are in the midrange of temperature, and blue stars are hot.

How big is space?

Space appears to be infinitely large, meaning that it goes on without end. There is a theory that space is curved, and that if we set out in one direction we would eventually return home, but it would take an infinite amount of time to complete this journey (see chapter 14).

How cold is space?

In space, where there are few atoms or molecules to provide us with warmth, the temperature is 3 degrees Centigrade about what we call absolute 0. That temperature is −270 degrees Centigrade, or −451 degrees Fahrenheit. That 3 degrees represents the heat left over from the formation of the Universe.

48. *Miranda Lebofsky contemplates the sky with the telescope known as Miranda.*

QUESTIONS ABOUT LIFE

What killed the dinosaurs? Can the same thing happen to us?

Children are led to the large and the vast, and the demise of the dinosaurs is a problem that brings in large animals, distant times, and great collisions. As we discussed in chapter 12, it is certain that a large comet or asteroid struck the Earth 65 million years ago. It is likely that the effects of this catastrophe resulted in the deaths of some three-quarters of all the species of life on Earth. This can be a difficult story to tell, for it arouses fear that we could be next.

It is important to express the statistics in their proper context. The dinosaur comet, most scientists think, was the cause of a very rare event that occurs maybe once in 100 million years. That does not mean that in 35 million years another comet will hit the Earth. What it does mean is that the chances that such an event will occur this year are only 1 in 100 million—a virtually impossible chance. This means that children can get interested in the event, but not have to fear it. Smaller comets, such as Shoemaker–Levy 9, could hit the Earth more often, like once in 100,000 to 500,000 years. If a person lives for a century, that means there is about a 1 in 1,000 chance that a comet or asteroid about a kilometer in diameter could hit the Earth during a long lifetime. That is a more distinct possibility, but still just one to be aware of and interested in, not one to be afraid of in this era of impact thriller movies.

Has life ever been detected on other worlds?

Not yet. Life has not been detected on any other planet. From the time of Antoniadi, whose late-nineteenth-century drawings of Martian channels suggested to Percival Lowell that a civilization might exist there, to the flyby of Mariner 4 in 1965, there was thought that life could exist there. There is still a *remote* possibility that simple life forms could exist there, or even under the ice of Jupiter's moon Europa. In 1996, scientists examining a rock from Mars found evidence that submicroscopic life might have existed on Mars a billion or more years ago (see chapters 9 and 16).

How come there isn't life on other planets?

A common question from children, to which we have to answer that we do not know that there is not life on other worlds. We have not found any such life, but we might some day (see chapter 16).

Do you believe in UFOs?

Far from being a flying saucer inhabited by aliens, a UFO is simply an unidentified flying object. Technically it means anything that the observer can't identify. In their years of observing the night sky, the authors of this book have not really seen anything they couldn't scientifically or logically explain.

As for seeing UFOs in the sense of flying saucers and aliens, we decidedly say no. As scientists, we try to apply the scientific method to any strange sighting that we see. What could it be? How does this sighting follow the laws of physics? We also follow the teaching of *Occam's Razor*. The simplest explanation of a phenomenon is likely to be the correct one.

One night we heard a story from someone who saw a ball of light roll across the ground and then take off, turn right, then suddenly speed away at 25,000 miles an hour. We didn't believe him. How could he determine that it launched itself that fast? What device did he use to measure its velocity? Such a sighting does not make sense, and while there are definitely strange goings-on in Nature, explanations that do not follow physical laws should be suspect (see chapter 13).

UNUSUAL QUESTIONS

How much do you make?

For astronomers visiting classrooms, this is a common question. Especially in the upper elementary and high school grades, children want to know. We usually say that astronomers, as most scientists, do not make an overly large amount of money, and that we do not become astronomers to get rich.

Are you married?

Children try to see how close their guest speaker is to being a "real person." One day Levy was asked this question while lecturing to a group in the school in which his girlfriend, Wendee Wallach, taught. A fellow teacher sitting next to Wendee quipped, "We're working on it!" (The answer to the question is now, "Yes.")

Can you get pregnant during an eclipse?

Questions like this are based on what people used to call "old wives' tales"; now we would call them tall tales or something like that. The day before the annular eclipse in 1994, an eighth-grade student in Las Cruces, New Mexico, asked this question. The answer, of course, is that women can get pregnant during an eclipse but not because of an eclipse. We need to explain to children that line-ups of the planets cannot affect us in this way (see chapter 7 and 8).

When the Dipper is upside down, will it rain?

It seems incredible that weather predictions could ever be based on such a poor understanding of the motion of the stars about the pole. The Dipper is circumpolar from most of North America, which means that as the Earth rotates it circles the pole, sometimes appearing upside down and sometimes right side up. This is different from using the appearance of asterisms such as the Pleiades in the eastern evening sky as harbingers of the winter rainy season. Another thought is that the Dipper bowl is filled with water. The stars in the Dipper are far away, not joined together by anything except our own imaginations.

PHILOSOPHICAL QUESTIONS AND THOUGHTS

A sense of place: What is the world?

Many questions deal with children trying to understand their place in the universe—their relation to their environment. The idea that the Earth is a round sphere in space circling the Sun is difficult to understand, and many of their questions relate to this.

297

In chapter 9, we discussed several lines of evidence that the Earth is a planet. Day and night show that the Earth is a world that rotates, the seasons help to show that the Earth is tilted and that it orbits the Sun, and the nightly movement of the Moon, and its phases, show that it is a body orbiting the Earth. These are every-day experiences and can only be explained by the understanding that the Earth is a world.

The Universe and religion.

Relating children's religious and spiritual beliefs to science is one of the major challenges of bringing science into the classroom. Questions of science and religion are most likely to occur during any discussion of the origin of the Earth or the origin of the solar system, or the Universe.

In their many forms, questions relating science and religion are as difficult to answer as they are common to ask. There may not be a single correct answer, but we do have a few suggestions.

The origin of the Universe is a subject that can be enjoyed scientifically, without necessarily infringing on religious beliefs or affecting the spiritual needs and feelings of the children. When a child asks a question relating to God, heaven, or religion, Levy often answers this way: "Science tries to explain the how, where, and when of things we see in Nature," he often says. "Religion seeks to answer the why." This simple statement offers one ap-proach to the problem. It is a strategy that encourages the idea that science and religion are different ways of inquiring about our surroundings. This is a valid idea, one that we should be trying to encourage.

However, science and religion offer two very different ways of looking at the world. The scientific method, which we have used in every one of the demonstrations in this book, draws conclusions and provides increases in knowledge that are based on the things we observe in Nature. There are no dogmatic truths in science. For example, there is overwhelming evidence that the Earth is a planet orbiting the Sun. But if we were to discover even more overwhelming evidence that the Earth is in fact a flat field supported on the backs of four elephants standing atop a turtle, the scientific method would compel us to drop at once the idea that the Earth is a planet. Science works with empirical data; there is no faith involved.

Most religions, on the other hand, are based on faith. They seek to answer the ancient questions of why we are here on Earth, and who put us here. The answers provided form the dogma that have persisted for many generations.

All this background is intended to give you an idea of what these different philosophies entail. The basic similarity is that both science and religion do try to portray the world around us and to place humans within that world.

What was the Universe like when it was a single point, and what was it like in the few seconds after the Big Bang?

These are philosophical questions, and there is no real way of visualizing such an event. However, it is our experience that, especially in the elementary grades, if we consider that question at all, the children are likely to suggest that "the single point was God," or "the early universe was heaven." Such answers are not irrelevant, and we should not ignore them. The children are acknowledging the fact that we have left the realm of science.

Who created the solar system?

Although this is obviously a religious question, the children may not always see it that way. They may simply want to know how the solar system was put together. Within the context of "how vs. why" discussed earlier, we would get as quickly as possible into the question of how.

Why are we here?

We don't know. Again, our "how vs. why" might work here.

What is the scientific method?

Most teachers see something called the scientific method as a recipe for learning new things by experimenting. We begin with a purpose or problem, and then we have a list of materials, a procedure to follow, a set of observations to make, and finally conclusions we reach. This approach, made famous by generations of children participating in science fairs, does reveal the basic way of scientific inquiry. But it leaves out two important concepts. As astronomer Steve Edberg wrote, "by following the scientific method as it is actually practiced by working scientists, students will see that no problem is ever really solved. Rather, they will discover that research is a never-ending exploration into the complexities—and ultimate simplicity—of Nature.

"The scientific method is a circular process, wherein each step feeds back into the others. At the end of one level, the results may be used in either of the other two, or may have to wait for a minor or major breakthrough to continue. The scientific process is really one of

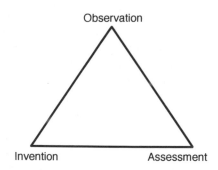

Edberg adds that assessment is often the weakest link in how we perceive the scientific method. "Children simply say their conclusion is right or wrong, without asking "How can I do this better?" or "Is there more to this problem that I am studying?"

Inspiration is the other aspect often left out of the scientific method. When chemist Friedrich Kekulé figured out the ring-shaped structure of benzene in 1865, he was inspired by a dream about small snakes. When he awoke, he took the intuitive leap from curling snakes to benzene rings. It is likely that very little scientific understanding would have been achieved over the ages had it not been for the sparks of feeling, imagination, and the desire to learn that led to the start of the process of understanding.

How do you become an astronomer?

This is a common question from primary school children. Our answer: In order to become an astronomer, you need to discover a star! Go outside one night and look up. Note how the stars are arranged. Find one bright star, then use a star chart to find out what star you have just seen. "There it is!" you might say. "It's Vega!" Then you have discovered a star. And then you are an astronomer.

Of course this dip into romance is not the real way to become a professional astronomer, a field for which you need all the mathematics and physics that you can get. For planetary astronomy, a good background in the geosciences also is important. But for elementary school children, this answer offers a way to launch them into careers of enjoying the night sky. At that stage, discovery is the most important thing.

Index

Index